INTERNATIONAL ENERGY AGENCY

# South American Gas

## Daring to Tap the Bounty

INTERNATIONAL ENERGY AGENCY
9, rue de la Fédération,
75739 Paris Cedex 15, France

ORGANISATION FOR
ECONOMIC CO-OPERATION
AND DEVELOPMENT

The International Energy Agency (IEA) is an autonomous body which was established in November 1974 within the framework of the Organisation for Economic Co-operation and Development (OECD) to implement an international energy programme.

It carries out a comprehensive programme of energy co-operation among twenty-six* of the OECD's thirty Member countries. The basic aims of the IEA are:

• to maintain and improve systems for coping with oil supply disruptions;

• to promote rational energy policies in a global context through co-operative relations with non-member countries, industry and international organisations;

• to operate a permanent information system on the international oil market;

• to improve the world's energy supply and demand structure by developing alternative energy sources and increasing the efficiency of energy use;

• to assist in the integration of environmental and energy policies.

* IEA Member countries: Australia, Austria, Belgium, Canada, the Czech Republic, Denmark, Finland, France, Germany, Greece, Hungary, Ireland, Italy, Japan, the Republic of Korea, Luxembourg, the Netherlands, New Zealand, Norway, Portugal, Spain, Sweden, Switzerland, Turkey, the United Kingdom, the United States. The European Commission also takes part in the work of the IEA.

Pursuant to Article 1 of the Convention signed in Paris on 14th December 1960, and which came into force on 30th September 1961, the Organisation for Economic Co-operation and Development (OECD) shall promote policies designed:

• to achieve the highest sustainable economic growth and employment and a rising standard of living in Member countries, while maintaining financial stability, and thus to contribute to the development of the world economy;

• to contribute to sound economic expansion in Member as well as non-member countries in the process of economic development; and

• to contribute to the expansion of world trade on a multilateral, non-discriminatory basis in accordance with international obligations.

The original Member countries of the OECD are Austria, Belgium, Canada, Denmark, France, Germany, Greece, Iceland, Ireland, Italy, Luxembourg, the Netherlands, Norway, Portugal, Spain, Sweden, Switzerland, Turkey, the United Kingdom and the United States. The following countries became Members subsequently through accession at the dates indicated hereafter: Japan (28th April 1964), Finland (28th January 1969), Australia (7th June 1971), New Zealand (29th May 1973), Mexico (18th May 1994), the Czech Republic (21st December 1995), Hungary (7th May 1996), Poland (22nd November 1996), the Republic of Korea (12th December 1996) and Slovakia (28th September 2000). The Commission of the European Communities takes part in the work of the OECD (Article 13 of the OECD Convention).

© OECD/IEA, 2003

# FOREWORD

South America has emerged in recent years as one of the most dynamic regions for natural gas. The continent boasts abundant gas reserves and high-growth energy markets. The need to diversify away from hydropower and oil is driving many countries to promote natural gas use, especially for power generation. Thanks to widespread democratisation and economic reforms that have opened to private investment a number of sectors previously reserved to the state, the region has been able to attract significant investment in exploration and production, gas processing plants, pipelines, LNG facilities and gas-fired power generation. Several large cross-border gas pipelines now link the countries of the Southern Cone. Many more are on the drawing board. In the north, Trinidad and Tobago is poised to become one of the largest LNG exporters in the Atlantic market.

This report reviews current trends in South America's natural gas supply, demand and emerging trade, with a particular focus on four major countries – Argentina, Bolivia, Brazil and Venezuela – that are, or are poised to become, major exporters or importers of natural gas. The report also analyses recent reforms in the structure and organisation of the region's gas industry and, in the light of the experience of other gas markets around the world, identifies the challenges ahead in order for the region to take full advantage of its gas resources.

The report raises some of the key policy issues that South American countries might need to address in order to foster gas-market development. The main message is that the development of natural gas markets is a long and challenging process, which requires sound energy and gas policies, credible institutions and a transparent and stable fiscal and regulatory framework. South American governments need to pursue efforts, both domestically and in co-ordination with their neighbours and trade partners, to identify and reduce barriers, and create the right incentives and guarantees that will mobilise domestic private resources and attract foreign investment.

The book is published under my authority as Executive Director of the International Energy Agency.

Robert Priddle
Executive Director

# ACKNOWLEDGEMENTS

The main author of this book is Sylvie D'Apote, of the IEA's Office of Non-Member Countries. Ralf Dickel, Head of the Energy Diversification Division, provided invaluable input and advice.

Comments and suggestions from several key South American and gas experts are gratefully acknowledged. In particular, the book benefited from the solid knowledge and through review of José Luis Aburto, Helder Pinto Junior and Gerardo Rabinovich.

Special thanks to other IEA colleagues who provided comments and support; in particular to Sylvie Cornot-Gandolphe for her detailed comments on gas markets, Anouk Honoré for assisting with the research and drafting; Pierpaolo Cazzola and Yukimi Shimura for helping with the statistical data, and Bertrand Sadin for the maps and graphics.

Christiane West and Miriam Oriolo provided useful input in the early stage of research.

Assistance with editing and preparation of the manuscript was provided by Christopher Henze, Christine Wallace and Scott Sullivan. Production assistance was provided by Loretta Ravera and Muriel Custodio.

# TABLE OF CONTENTS

## LIST OF TABLES

## LIST OF FIGURES

## LIST OF BOXES

## LIST OF MAPS

# EXECUTIVE SUMMARY

## OVERVIEW OF SOUTH AMERICAN GAS SUPPLY AND DEMAND

South America has emerged in recent years as one of the fastest growing markets for natural gas, attracting significant investment in exploration and production, gas processing plants, pipelines, LNG facilities and gas-fired power generation. Widespread structural reforms and efforts towards macroeconomic stabilisation helped South American countries recover sustained levels of economic growth in the 1990s, after a "lost decade", as the 1980s have been called. Hand in hand with rapid economic growth came also a rapid increase in regional energy demand. In most South American countries, energy demand grew faster than GDP. Despite recent political and economic instability in a number of countries, energy demand is still growing strongly, generally continuing its upward trend even in years of recession.

As in other parts of the world, natural gas is the fastest-growing primary fuel in South America. Over the past decade, gas demand in South America increased by 5.1% per year, while total energy demand grew at a rate of 3.2%. In Brazil and Chile, which used little gas in 1990, gas demand grew by 12% and 14% respectively from 1990 to 2000. As a result, the share of gas in South America's total primary energy supply increased from 18% in 1990 to 22% in 2000, comparable to the share of gas in North America and Europe.

**Gas reserves are abundant...** The continent boasts abundant natural gas resources, which are in large part associated with oil. Proven gas reserves have increased by 50% since 1990 and nearly tripled since 1980. They were above 7 tcm at the beginning of 2002. Probable and possible reserves could add another 5 tcm. Recent large discoveries in Bolivia and offshore Trinidad & Tobago suggest that intensified exploration could lead to a substantial increase in the region's proven reserves in the medium term. In addition, Brazil's deep offshore fields are thought to have a large potential.

While regional gas resources are plentiful, they are not generally situated within easy reach of potential regional markets. Long distances and formidable geographical obstacles (the Andes and the Amazon forest) make the transportation of South America's gas resources to the region's main population and industrial centres a difficult and costly venture.

Seventy percent of the region's gas reserves are situated in Venezuela and Trinidad & Tobago, at the far north of the continent. While these gas resources are well-located to supply the North Atlantic markets, they are separated from the southern half of South America, where 70% of the continent's population lives, by several thousand kilometres of impenetrable jungles and sparsely populated areas. Other gas reserves have been found in remote areas such as the southern tip of Argentina, the Peruvian jungle and the Brazilian Amazon. Some of these reserves are in difficult terrain or in environmentally sensitive areas. Many other isolated and little-explored areas are thought to hold substantial gas resources.

**...but lack of markets is the limiting factor**

While the region has well-developed oil and hydropower resources, it is really only just starting to develop its gas resources. The experience of the last ten years has shown that the main limiting factors in South America are not reserves and production, but markets and infrastructure. Argentina is the only country in the region with both a high level of domestic gas reserves and a mature gas market, with a well-developed infrastructure and a high level of gas penetration – comparable to that of Russia and the Netherlands. Bolivia, Peru and Trinidad & Tobago have large gas reserves, but small and limited domestic gas markets. On the other hand, Brazil and Chile, where annual gas demand is growing at two-digit percentage rates, depend mainly on imports. Venezuela and Colombia, oil producers with significant associated-gas reserves, are keen to increase their domestic gas consumption, but gas production is dependent on oil production, and it is doubtful that their domestic gas markets alone will provide enough opportunities and incentives to spur exploration and production of non-associated gas.

**Industry is the largest gas user**

Patterns of gas consumption differ greatly from country to country. Apart from Argentina, which has a high level of gas penetration in all end-use sectors, elsewhere on the continent the use of gas has traditionally been limited to a few oil-producing countries, where gas discovered as a result of oil exploration was used mainly in the industrial sector and in the oil and gas sector itself. Currently, industry is the largest gas user in South America.

Gas demand in the residential, commercial and public services sectors is limited because there is no need for space heating in most of the continent. The use of gas for cooling may provide opportunities to expand gas use in the commercial and public sector. The use of compressed natural gas (CNG) as a transport fuel is expanding rapidly: Argentina is the world leader in this area, with nearly 300,000 CNG vehicles, and other South American countries are following suit.

**Gas-fired generation is still low, but growing**

Gas use for power generation is still low at the regional level and is concentrated in just a few countries. This is because the region has large and well-developed hydropower resources. In 2000, hydropower provided three-quarters of the region's electricity supply, a much higher share than in any other region of the world. While there is still considerable potential for hydroelectric expansion in several countries, public budgets can no longer afford to finance large infrastructure projects, and private investors prefer gas-fired power stations because they have lower up-front costs and shorter lead times. In addition, recent droughts in Argentina, Brazil, Chile and Venezuela have exposed the vulnerability that comes with high dependence on hydropower. Several countries are actively promoting gas-fired thermal generation to increase diversification of their power-generation mix. For all these reasons, much of the new capacity built over the past five years has been gas-fired. Gas-fired generation grew by 8.4% per year between 1995 and 2000, compared with 4.5% for total generation.

**Gas markets are increasingly interconnected in the Southern Cone**

Given the uneven distribution of reserves and markets, gas projects in South America in most cases involve cross-border trade of either pipeline gas or LNG. The development of gas pipeline interconnections is most advanced in the Southern Cone, a region encompassing the southern half of Brazil, Argentina, Chile, Bolivia, Paraguay and Uruguay. This is where most of the population and industrial infrastructure is located and where the growth in energy and gas demand is highest. Both Argentina and Bolivia have abundant non-associated gas reserves, which they are eager to export to neighbouring countries.

Between 1996 and 2001, seven pipelines were built between Argentina and Chile, the 3,150-km Bolivia-to-Brazil pipeline was finalised and the first stage of an Argentina-to-Brazil pipeline became operational, laying the basis for a sub-regional gas transportation network. Another pipeline from Argentina to southern Brazil via Uruguay is at an advanced planning stage and a possible route from southern Bolivia to Brazil via Argentina and Paraguay is being considered. In 2001, gas trade in the Southern Cone amounted to 9 bcm, or 16% of the Southern Cone's marketed gas production.

Gas pipeline interconnections among the Andean countries – Bolivia, Peru, Ecuador, Colombia and Venezuela – are likely to progress much more slowly because, unlike in the Southern Cone, there are no evident energy complementarities between these countries, which have oil and gas resources in varying degrees. The only project currently under study is a gas link between Colombia and Venezuela.

**LNG exports are stepping up on the Caribbean coast and planned on the Pacific**

While the large reserves in the north are too far away to supply pipeline gas to the Southern Cone markets, they offer great potential for LNG projects. Trinidad & Tobago inaugurated its first LNG train in 2000, and a second one in 2002. In 2001, it exported 3.8 bcm to the US East Coast and Spain. There are plans to add three more LNG trains by 2005, which would make Trinidad & Tobago one of the largest LNG suppliers of the Atlantic market. Venezuela certainly has enough gas reserves to become a major LNG exporter while meeting greatly increased domestic demand. Projects for LNG liquefaction plants in Venezuela have long been stalled by a combination of poor economics and lack of political support. Technological developments, which are lowering the costs in the LNG chain, as well as a new policy favouring gas projects to reduce the country's dependence on oil exports, are giving new impetus to LNG projects. The government has recently awarded several concessions to develop offshore gas reserves, whose output will be used partly in LNG export projects.

Further south, and looking at the Pacific market, Bolivia is seeking to capitalise on its enormous gas reserves and is exploring the possibility of exporting LNG to the west coast of Mexico and the US, via a pipeline to a port in Chile or in Peru, where a liquefaction plant would be built. Peruvian gas from the giant Camisea field may also one day be exported as LNG, since the local market is very small and pipeline exports to Brazil now seem precluded by the abundance of gas in Bolivia. South America will likely become a significant LNG exporter in a not too distant future.

As yet, LNG is not imported. However, there are plans to construct an LNG importing terminal on the eastern coast of Brazil. LNG could come from Trinidad & Tobago, Nigeria or, eventually, Venezuela.

**A decade of market reforms and privatisation...**

Whilst substantial gas reserves exist and the potential for market development is considerable, the investments needed to bring projects to fruition are enormous. In the last decade, South America has emerged as a very attractive region for private investment, thanks to the democratisation and economic liberalisation processes that have opened to private investment a number of sectors previously reserved to the state.

Argentina was the first to launch thorough reforms of its oil and gas industry in the early 1990s, with the unbundling and privatisation of its state-owned *Yacimientos Petrolíferos Fiscales* (YPF) and *Gas del Estado* (GdE). The Argentine gas market is currently the most liberalised in the region, with private companies in all segments of the industry. Peru and Bolivia followed suit in the mid-1990s. In Venezuela, Colombia, Brazil, Chile, Ecuador and Trinidad & Tobago, oil and gas companies remain in public hands, but reforms have been carried out to allow or increase private participation in some or all links of the gas chain. In the downstream sector, most of these countries have moved towards liberalisation of gas transmission and distribution. Venezuela is the only country where these activities are still 100% in the hands of a state-controlled company, but even in Venezuela there are now plans to award concessions to private companies for the construction and/or operation of pipelines and distribution networks.

The opening of the upstream sector to a larger number of companies has in many cases helped to firm up gas reserve estimates, boost production and lower the development and production costs of domestic gas. Bolivia and Trinidad & Tobago are the best examples of this trend. In Bolivia, the large infusion of private capital in the upstream sector brought a surge of investment in exploration and production, which produced a seven-fold increase in proven gas reserves in just four years. Through the involvement of the private sector, the region has also been able to attract the necessary capital and technology to undertake successfully a number of large gas pipeline projects that only a few years ago would have seemed impossible.

Gas-sector reforms in South America have generally been part of wider energy-sector reforms, with substantial power-sector reforms taking place before or simultaneously. Indeed, Chile and Argentina were at the forefront of electricity reforms during the 1980s and early 1990s, providing an impulse for changes in neighbouring countries.

**...has opened the door to many new entrants...**

Market transformation and liberalisation have provided opportunities not only for new players to enter the South America energy arena, but also for traditional players to broaden their sphere of action. In the upstream oil and gas sector, the incorporation of private players was only in a few cases (Argentina, Bolivia, Peru) the result of the break-up and sale of state companies. In general, this process has involved the participation of foreign companies, sometimes in partnership with regional private companies. In the rest of South America, while state companies remain dominant, and in many cases are granted substantial prerogatives, legal reforms have allowed private players in activities previously reserved to state monopolies or characterised by strong entry barriers. Here, new entrants were rarely new to the market; in most cases they were regional or foreign private companies which had been developing activities in niche or non-restricted sectors, such as the production of marginal fields, oil services, or construction and equipment manufacturing. In many cases, there has been a consolidation along the gas chain with upstream oil and gas companies acquiring stakes in gas transportation and distribution assets and in power generation.

Market players in South America now include, in addition to the American and European majors found in the oil and gas sectors throughout the world, other smaller European and North American companies and a significant number of private local companies, most of which are from Argentina. Some of largest foreign players in the South American gas market are Repsol-YPF, BP Amoco, TotalFinaElf, Shell, BG Group, and ExxonMobil. In many cases, the establishment of consortia between foreign and local companies has allowed the former to contribute technology and/or capital and the latter to provide in-depth country or regional experience.

**...and forced state companies to change**

The state companies, especially the largest ones, such as Venezuela's PDVSA and Brazil's Petrobras, have adapted remarkably well to the new national and regional environment, modifying their behaviour and undergoing significant transformations (restructuring, reorganisation, changes in managerial practices, partial break-up of assets, refocusing on core activities, etc). These companies now have different objectives, strategies and tools, allowing them to seize business opportunities, not only on their national territory but regionally and internationally as well. Petrobras, in particular, has large gas reserves in Bolivia and has decided to acquire stakes in Argentina's Perez Companc, which accounts for 5% of Argentina's gas reserves and production.

**Regional co-operation and integration are advancing, especially in the Southern Cone**

Because many gas projects in South America are cross-border, a degree of regional co-operation and harmonisation is essential. The development of regional trade blocks such as Mercosur and the Andean Community has not only facilitated cross-border trade through the gradual elimination of import/export duties, but has also fostered greater political stability and economic growth throughout the region, resulting in higher energy demand growth.

While the countries of the Andean Community have relatively isolated and self-sufficient energy systems, except for a few localised electricity interconnections, Mercosur countries are undergoing rapid physical energy integration and are working towards establishing a minimum of compatibility in national fiscal and regulatory frameworks (including agreements to prevent double taxation) to facilitate cross-border projects and energy transactions between companies of different countries. It is noteworthy that in several instances the construction of cross-border pipelines and the establishment of common commercial energy interests have helped to tone down historical frictions between neighbouring countries. This is certainly the case for Argentina and Chile and, more recently, for Bolivia and Chile.

## WHAT PROSPECTS FOR THE NEXT DECADE?

Clearly, there is substantial potential for increased gas production, consumption and trade in South America. Regional energy demand is growing at rates above the world average. This trend is expected to continue in the next decade, despite a likely slowdown due to the Argentine crisis and the slowing of the US economy.

**The power sector is expected to drive gas demand growth**

Future gas demand in South America is expected to be driven largely by the increase in gas use for power generation. As in other parts of the world, gas-fired CCGT is the preferred technology for new power projects, due to its lower up-front costs, shorter lead times, high efficiency and flexibility, and low environmental impacts. In addition, South American countries with a high dependence on hydropower are seeking to promote gas-fired generation in order to diversify their power mix. Furthermore, many countries see gas-fired power plants as the necessary first step to develop their gas markets, as power plants provide the necessary anchor demand to ensure the viability of investments in gas production and infrastructure. However, in hydro-dominated power systems, such as in Brazil, the competition from hydropower plants, in large part amortised, may harm the profitability of gas-fired power plants, thereby affecting the viability of the whole gas chain.

**Significant potential for expansion of gas use in the industrial sector**

There is also substantial scope for further substitution of gas for oil in the industrial sector. A few gas-producing South American countries still have little or no industrial gas use. Oil-exporting countries may also wish to promote the use of gas in industry, so that displaced oil can be exported. In oil importing countries, such as Chile, imported gas may displace more expensive imported oil, at the same time diversifying the energy mix and lessening dependence on oil imports. Moreover, in many South American countries, growing concerns about local air pollution and participation in international treaties limiting greenhouse gas emissions may motivate governments to enforce environmental legislation, encouraging further fuel-switching by industrial consumers and power plants. The key issues for the industrial sector are gas prices in relation to competing fuels, and environmental policies and regulations.

**Regional demand will drive interconnections in the Southern Cone...**

In the Southern Cone, gas developments will be driven by the expanding markets of Brazil and Chile. Supply will come from Bolivia and Argentina, with new fields expected to be brought into production and several new cross-border pipelines either planned or already under construction to link production with consumption centres. The Argentine economic crisis, and its effect on neighbouring countries, may slow this trend, but will not stop it. One of the main uncertainties surrounding the pace of development of the Southern Cone's gas reserves and transport infrastructure is future gas demand in the Brazilian power sector. In that country, regulatory uncertainties, gas pricing issues and the inherent complexity of introducing gas-fired combined-cycle plants in a hydro-dominated system are delaying thermal power projects with a total capacity of several thousand megawatts.

**...while LNG exports will drive reserve development in the North**

In the northern part of the continent, reserve development will be driven by the potential for LNG exports to distant markets, which are also expected to show high growth in demand. Some South American LNG will continue to flow to Europe. But the main market for South American LNG, from both the Caribbean and the Pacific coast, is likely to remain North America, especially the United States. Most observers agree that the United States' traditional sources of gas – domestic production and imports from Canada – will not be able to keep up with demand, which is expected to increase substantially over the next decade, particularly for power generation. Some South

American LNG is also likely to be imported by Mexico, either for re-export to the US or to supply its own rapidly-growing domestic market, which is quickly outstripping production capacity. There are several projects in Mexico and in the US to build new LNG terminals or reactivate terminals which had been mothballed.

For LNG projects, the main uncertainty is related to demand and, more importantly, to gas prices in the United States, which have proven difficult to predict in the past, and which will determine the profitability and economic viability of South American LNG projects. Another factor is the competition from other existing and projected LNG projects, such as in Nigeria for the Atlantic market and in Alaska, the Canadian Arctic and even Australia for the Pacific market.

An alternative way to commercialise gas reserves that are far from markets in South America and without convenient access to the coast is through gas-to-liquid (GTL) schemes. The competitiveness of GTL versus LNG will depend on further technological developments to reduce costs of GTL and on the individual circumstances of each project.

## MEETING THE CHALLENGE

While South America is one of the world's most promising zones for the development of natural gas, fulfilling the promise will depend largely on creating the right environment to win investors' confidence. State budgets and state-owned companies can no longer finance large energy projects. Hence, the ability to attract private capital will be crucial for South America to take full advantage of its gas bounty.

The development of natural gas markets is a long and challenging process, which requires sound energy and gas policies, credible institutions and a transparent and stable fiscal and regulatory framework. Some South American countries have already gone some distance in that direction. Others are just starting the process. South American governments need to pursue efforts, both domestically and in co-ordination with their neighbours and trade partners, to identify and reduce barriers, and create the right incentives and guarantees that will mobilise domestic private resources and attract foreign investment.

The opportunities and challenges for expanding the gas sectors in South America vary as much as the countries themselves. It is beyond the scope of this report to recommend a tailored solution for each country. Instead, it aims to raise some of the key policy issues that South American countries might need to address in order to foster gas-market development.

**Setting a comprehensive approach to energy policy**

As gas markets move towards increased private-sector participation, the role of government shifts but does not disappear. While commercial decisions on gas supply allocation and prices and detailed investment choices are best left to the market, policy formulation remains a fundamental role of the government. Gas-sector policy should be part of a coherent and integrated energy-sector strategy, with particular attention given to the proper design of electricity markets, as gas-to-power projects are the key to ensuring the financial viability of the whole gas chain. Co-ordination between the energy ministry and other government agencies is also important because energy policy interfaces with many environmental, fiscal, social and industrial policy objectives.

**Strengthening policy-making capacity**

In view of the complex issues involved, governments should strive to establish and maintain a high level of policy-formulation capacity. Energy ministries should be given sufficient resources and maintain an adequate level of technical expertise in order to play an effective role in defining their countries' energy policy. They should monitor energy developments closely so that they are able to react quickly to market failures and other crises, without relying unduly on the regulatory agencies, whose task is to implement – not formulate – government policies.

**The importance of pricing and taxation**

Gas pricing and taxation are especially important in emerging gas markets, as gas competes with other fuels. Pricing decisions are best left to market players, with the state's role restricted to preventing anti-competitive behaviour and discrimination. In countries which do not have large and inexpensive gas reserves close to demand centres, a sound gas pricing system should start with the final consumers' willingness to pay, determined by the overall costs of using alternative fuels. This requires that the prices of competing fuels should also be market-based and undistorted.

Taxation of natural gas and other fuels is a key element of a coherent energy policy, and can play a critical role in stimulating the supply of and demand for natural gas, especially in the early stages of the industry's development. Favourable taxation of gas *vis-à-vis* alternative fuels should be used to internalise externalities and to provide economic signals to the market as an instrument to implement government policy.

**Minimising the country risk and reducing legal and regulatory uncertainty**

Because of their need for high up-front investment, long lead times and because of their high specificity, gas projects are particularly sensitive to risk, as risk increases financing costs. Sound economic policies and financial stability will reduce the country risk. Clear, transparent and predictable legislation and regulation will also contribute to reassuring investors and financing institutions. Separating responsibilities for policy making, regulation and – if applicable – dealing with state companies is key to improving transparency and reducing the arbitrariness of decision-making, ensuring a level playing field for all participants.

**Promoting cross-border co-operation and harmonisation**

Cross-border gas projects do not necessarily need full harmonisation of regulation or similar degrees of market liberalisation on each side of the border, although these may be worthwhile long-term objectives. However, they need clear policies and stable legal and regulatory frameworks in each of the countries involved, as well as some degree of co-ordination between countries, for example to avoid double taxation.

## STRUCTURE OF THE REPORT

This report is divided into two parts, following a brief *Introduction* describing how gas industries and markets differ from those of other fossil fuels and electricity.

*Part I* reviews current trends in South America's gas sector and identifies the challenges ahead for the region to take full advantage of its gas resources.

■ *Chapter 1* discusses current and expected trends in gas supply in South America, examining the type and location of reserves and the level of production.

■ *Chapter 2* analyses South America's existing and potential gas markets, highlighting the main engines of growth in terms of geographical areas and sectors.

■ *Chapter 3* looks at South America's emerging gas trade. It reviews existing and projected cross-border pipelines and assesses the potential for LNG exports.

■ *Chapter 4* looks at the recent reforms in the organisation and structure of the gas industry in South America. In the light of experience in other gas markets around the world, it identifies the main challenges facing South American countries today if they are to benefit fully from their gas resources.

*Part II* comprises individual profiles of Argentina, Bolivia, Brazil and Venezuela. These countries account for 80% of the region's natural gas proven reserves and production.

■ *Argentina* is a net gas exporter and boasts the most mature gas market in the continent. A leader in the privatisation of state-owned utilities, Argentina implemented sweeping changes to its upstream and downstream gas industry in the early 1990s.

■ *Bolivia*'s geographical position and recent remarkable increase in gas reserves make it ideally placed to become South America's gas hub and play a central role in Southern Cone energy integration.

■ *Brazil* is the largest energy market in South America. Despite current low gas penetration, its gas demand and imports are expected to increase significantly in the near future, driving reserve development in neighbouring producing countries.

■ *Venezuela* holds the eighth-largest gas reserves in the world, mostly associated with oil. Though its gas production is still comparatively low, the government is now pushing to develop non-associated gas reserves and increase domestic gas consumption and exports.

# INTRODUCTION

## THE SPECIFICITIES OF NATURAL GAS INDUSTRIES AND MARKETS

**Gas is different from other fossil fuels**

Natural gas is significantly different from other fossil fuels because of its low energy density and its gaseous state. Unlike oil and coal, natural gas cannot be inexpensively stored in large quantities, and sophisticated and expensive infrastructure is required to deliver gas to the end-user. Hence, gas transportation costs are very high. In general, transportation represents 50% of the total final consumer price, compared to 5-10% in the case of oil. Additionally, gas transportation and distribution systems involve large economies of scale and are relatively inflexible; once the infrastructure is built, it cannot be used for other purposes. For these reasons, small-scale gas markets are difficult to develop.

Unlike oil and coal, there is no world gas market: 85% of the gas consumed in the world today is produced locally, and gas trade occurs largely on a regional scale. Thus, there is no world standard for gas pricing or marker price, such as WTI or Brent for crude oil. Small and remote gas discoveries are often hard to commercialise. As a result, there is a large amount of "stranded gas" in the world.

The most significant difference is that, while investment in oil and coal production can be made independently from the conclusion of sales contracts, relying on selling the product on the world market, investment in gas production cannot proceed without long-term commitments between prospective producers and consumers.

**Gas is different from electricity**

Although gas and electricity are both grid-bound industries, they have significant differences. The main difference is that gas is only produced where it is found, while electricity, except for hydropower, can be produced anywhere. Thus, transportation plays a different role for gas compared to electricity. Another difference is that, with a few exceptions, alternative fuels can always be substituted for gas, which is not the case for electricity. Thus, gas pricing must take due account of the competitiveness of gas versus other fuels.

**The gas industry works like a chain...**

The gas industry works like a chain linking the wellhead and processing plant to the transmission network, to the distribution network, and, finally, to the consumer. Each physical link corresponds to a commercial relationship. Each link is dependent on every other link, and the overall strength of the chain depends on its weakest link. As the whole network is vulnerable to disruption, firm and long-term relationships – in the form of "take-or-pay" contracts or vertical integration – are the most common.

**...it needs huge investments...**

The gas industry is very capital-intensive: huge up-front investments are required to produce, transport, store and distribute gas. At normal temperature and pressure, natural gas is about a thousand times more voluminous than oil for the same energy content. Even under high pressure, the energy density of natural gas is about one-tenth that of oil: this means that an oil pipeline can transport about ten times as much energy as a gas pipeline of the same diameter. Thus, transportation costs makes up a greater share of the final price of gas than is the case for other fuels.

**...in a co-ordinated manner**

The co-ordination of investments along the gas chain is crucial. In many cases, gas resources are far from potential demand centres and require long-distance transportation. Field development and pipeline construction have no value unless a market infrastructure such as gas-fired power plants and local distribution networks is created simultaneously.

**Gas-market development takes time...**

The development of natural gas markets is a long process. It took 50 years for the share of gas in the world's energy mix to increase from 2% in 1900 to 10% in 1950. During that period, more than 90% of the gas was consumed in the United States where, despite huge local resources and strong potential demand, the emergence of a gas market faced major challenges. In Western Europe, natural gas was not consumed until after World War II and reached only 20 bcm in 1965. Four decades later, European gas industries and markets are still in varying degrees of development.

**...and is driven by paying demand**

The last, but perhaps most important, point is that, unlike oil and coal, the development of natural gas is driven by demand, rather than by production, more precisely by a demand that can justify financing for the investment required all along the gas chain. Resources and infrastructure alone are not sufficient; the market for end-uses must be developed. Moreover, unlike electricity, gas must compete with other energy sources. For example, in the residential and commercial sectors, gas competes with electricity, liquefied petroleum gas (LPG) and heating oil. In the industrial sector, gas competes with coal and oil for use in steam boilers. In the production of electricity, gas competes with coal, oil, nuclear, hydro, biomass and other renewables.

**Special role of the state in the natural gas industry**

Gas production uses non-renewable natural resources that are part of a country's national assets. Gas transportation and distribution require a fixed infrastructure that uses public land and public streets. All gas activities have important health, safety and environmental consequences. Furthermore, gas distribution and, in some cases, transmission have the character of a natural monopoly. In addition, the power sector, which is a main driver for the development of gas markets, requires clear government policy in order to perform effectively and provide the revenue needed to sustain the gas sector.

For the above reasons, the state – either the central government or the local municipalities – has an important role to play. In the past, in most countries the state managed gas activities directly, through state-owned companies. This, however, has changed. The current trend is to let commercial players handle the commercial activities of the gas industry as much as possible. Nevertheless, the role of the government remains essential in establishing a clear policy, regulatory and fiscal framework for the gas industry. This framework should stimulate investment, protect consumer interests and provide a fair revenue for the state. This is particularly important at the early stage of gas-market development.

# PART I

OVERVIEW OF NATURAL GAS MARKETS IN SOUTH AMERICA

# CHAPTER 1
# NATURAL GAS SUPPLY

## NATURAL GAS RESERVES

**Abundant and relatively widespread**

Proven natural gas reserves in South America[1] amounted to 7.1 trillion cubic metres (tcm) at the beginning of 2002, of which nearly 60% are located in Venezuela. They represent 4% of global gas reserves. *Map 1*[2] shows the location of the region's main gas reserves, while *Map 2* gives an overview of each country's endowment and production.

**Table 1.1** Proven natural gas reserves in South America, 1980-2002 (billion cubic metres)

| As of 1 January | 1980 | 1990 | 2000 | 2001 | 2002 | % |
|---|---|---|---|---|---|---|
| Argentina | 574 | 671 | 748 | 778 | 764 | 11% |
| Bolivia | 130 | 116 | 518 | 675 | 775 | 11% |
| Brazil | 45 | 116 | 231 | 221 | 220 | 3% |
| Chile | 129 | 115 | 95 | 94 | 93 | 1% |
| Colombia | 134 | 112 | 195 | 198 | 198 | 3% |
| Ecuador | 120 | 112 | 103 | 104 | 115 | 2% |
| Peru | 37 | 340 | 255 | 255 | 255 | 4% |
| Trinidad & Tobago | 200 | 286 | 602 | 705 | 558 | 8% |
| Venezuela | 1,249 | 2,993 | 4,148 | 4,163 | 4,163 | 58% |
| **South America** | **2,618** | **4,861** | **6,896** | **7,193** | **7,140** | 100% |
| **World** | **76,871** | **129,318** | **159,040** | **164,745** | **178,270** | |

*Note: The decrease in the reserves of Trinidad & Tobago is due to a change of methodology, not to a real decrease of reserves.*
*Sources: Countries' official information, OLADE, Cedigaz.*

South America has experienced an unprecedented growth in proven natural gas reserves in recent years. Proven natural gas reserves have increased by 50% since 1990 and have nearly tripled since 1980. This is equivalent to an average annual increase of 4.7% in the past two decades, well above the world's average of 3.9%. Aggressive exploration spurred by expected demand growth has yielded large new gas discoveries, and petroleum companies have stepped up the development of existing fields. In the Southern Cone, fears that reserves would not be sufficient to meet large increases in demand, and thus

---

1.  For the purpose of this report, South America includes Argentina, Bolivia, Brazil, Chile, Colombia, Ecuador, Paraguay, Peru, Trinidad & Tobago, Uruguay and Venezuela. Thus, total figures for South America do not include French Guiana, Guyana, and Suriname, unless specified.
2.  All maps are located at the end of the book.

not justify the substantial investments required for pipelines, have proved to be unfounded. Indeed, the opening of new markets, as in the case of the Bolivia-Brazil pipeline, has created an incentive for intensified exploration.

**Associated and non-associated gas reserves**

An important distinction, which affects the current and potential exploitation of gas reserves, is whether those reserves are found in reservoirs that contain mostly gaseous hydrocarbons (non-associated or free gas) or in reservoirs that hold a large percentage of liquid hydrocarbons (associated gas). Despite significant recent discoveries of non-associated gas (notably in Bolivia and Peru), associated gas still accounts for over half of proven gas reserves in South America, and the share is as high as 91% in Venezuela.

This reflects the fact that much gas in South America was found as a result of oil exploration. Still today, in many South American countries, oil production is much higher than gas production (*Figure 1.1*). Usually, the economic benefits of producing oil are much greater than for gas. The infrastructure necessary to process and transport gas is costly and requires long lead times. Hence, oil companies generally view associated gas as an undesirable by-product: in order to make use of it, they need to invest more and defer oil production. When associated gas is found in fields too remote from demand centres, or when potential gas demand is too small to justify heavy investments in gas infrastructure, the gas is either reinjected into the reservoir, flared or, in some cases, vented.

Furthermore, where the share of associated gas in total reserves is high, production of gas is highly dependent on oil production, and hence on the demand for, and the price of, crude oil. This is especially visible in Venezuela, where 100% of gas production is associated with oil and where the impact of OPEC quotas on oil production has precluded a steady supply of associated gas, limiting the development of the gas market. These

**Figure 1.1** Production of oil and gas by country, 2001

| | | | | | Mtoe | | | | | | |
|---|---|---|---|---|---|---|---|---|---|---|---|
| **Crude oil** | 40.0 | 1.8 | 66.3 | 0.3 | 31.3 | 22.1 | 4.6 | 6.2 | 169.4 | | 342.0 |
| **NGL** | 0.6 | 0.3 | 1.3 | 0.1 | 0.5 | 0.0 | 0.2 | 0.7 | 7.1 | | 108 |
| **Gas** | 42.9 | 67 | 13.5 | 2.8 | 11.8 | 1.3 | 0.7 | 162 | 66.4 | | 162.4 |

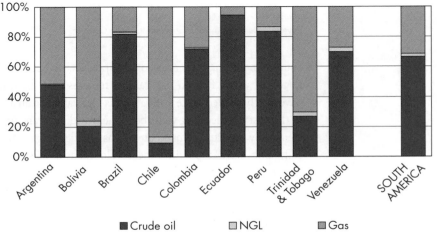

Sources: IEA, Cedigaz.

constraints are encouraging exploration and production of non-associated gas resources. While proven gas reserves are largely associated with oil, in most South American countries significant non-associated gas reserves remain to be proven and discovered.

## NATURAL GAS PRODUCTION

**Gross gas production is increasing...**

South America's gross production of natural gas reached 167 bcm in 2001 up 3.5% from 2000.[3] Gross gas production increased 5.5% annually between 1991 and 2001, a substantial acceleration from the 3.0% average annual growth of the previous decade. Of total regional gas production in 2001, Venezuela accounted for 37%, Argentina for 28%, Trinidad & Tobago for 10%, and Brazil and Colombia for 9% each. As shown in *Table 1.2,* in the last ten years, gross gas production grew fastest in Colombia (10.5% per year), Brazil (8.2%), Trinidad & Tobago (7.5%) and Argentina (7.1%).

**Table 1.2** Gross natural gas production by country, 2001

| | Gross production | | Aver. annual growth | | Reinjected | Flared & vented | Other losses |
|---|---|---|---|---|---|---|---|
| | bcm | % | 1991 -2001 | 1981 -1991 | % of gross production | | |
| Argentina | 45.9 | 28% | 7.1% | 5.5% | 7% | 1% | 11% |
| Bolivia | 7.2 | 4% | 2.8% | 0.9% | 32% | 8% | 4% |
| Brazil | 14.6 | 9% | 8.2% | 10.3% | 19% | 18% | 22% |
| Chile | 3.0 | 2% | -3.4% | -1.9% | 54% | 3% | 2% |
| Colombia | 14.2 | 9% | 10.5% | 1.0% | 54% | 4% | 2% |
| Ecuador | 1.2 | 1% | 5.6% | 5.6% | 17% | 70% | 0% |
| Peru | 0.9 | 1% | -4.3% | -4.5% | 40% | 19% | 0% |
| Trinidad & Tobago | 17.4 | 10% | 7.5% | 5.4% | 0% | 13% | 0% |
| Venezuela | 62.4 | 37% | 4.0% | 2.2% | 34% | 5% | 10% |
| **South America** | **166.6** | **100%** | **5.5%** | **3.0%** | **24%** | **6%** | **9%** |

*Source: Cedigaz.*

**...but reinjection and flaring is still high...**

In 2001, about 24% of gross production was reinjected into reservoirs, and 6% was flared or released into the atmosphere (*Figure 1.2*). Other losses or "volume shrinkage" account for a further 9%. Volume shrinkage is the result of the processing and treatment of natural gas before it enters the network. This is done to extract natural gas's liquid fractions (liquid petroleum gas, natural gasoline and condensates) and to remove impurities, such as water, acid gases, nitrogen, helium or mercury. These natural gas liquids (NGL) are valuable by-products of natural gas production; therefore extraction and commercialisation of NGL can greatly enhance the economics of natural gas projects. Indeed, according to Cedigaz data, volume shrinkage (in large part due to NGL extraction) has increased gradually from 3% of gross production in 1975 to 9% in 2001. NGL extraction is likely to continue to increase as more processing facilities are built.

---

3. Data for gross gas production, reinjection, flared/vented gas and "other losses" are from Cedigaz (2002). For definitions of gross and marketed gas production, and other terms and statistical aggregates used in this report, see Glossary.

In some cases, reinjection of gas is necessary to maintain pressure in mature oil reservoirs and boost declining oil production. This is the case, for example, in Venezuela and Colombia, where respectively 34% and 54% of total gross gas production are reinjected. The proportion of gas reinjected or flared is even higher in Ecuador and Peru, where oil and gas wells are deep in the jungle and the potential market for gas is very small. In Chile, gas production is mainly located in the far south of the country, a long way from the country's main consumption centres. In those countries, all gas that cannot be used locally is reinjected.

While gas reinjected into a reservoir can be extracted at a later date, flared or vented gas is lost forever. The practice of flaring and venting gas, however, has lessened considerably in recent years, due to pressures on the oil and gas industry to limit

**Figure 1.2** Gross and marketed natural gas production in South America, 1975-2001

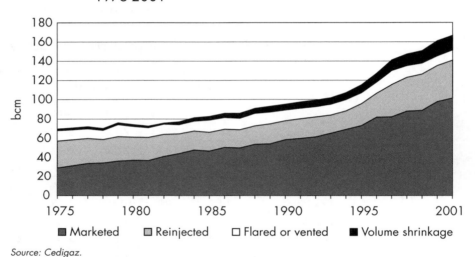

*Source: Cedigaz.*

**Figure 1.3** Gross and marketed natural gas production by country, 1975-2001

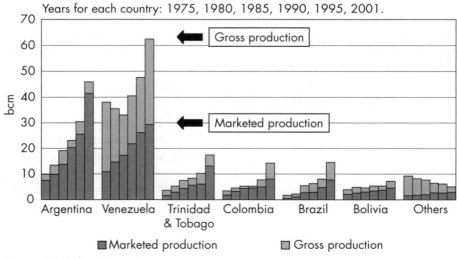

*Sources: IEA, Cedigaz.*

greenhouse gas emissions.[4] The share of gas flared or vented in the whole region decreased from 32% in 1970 to 6% in 2001.

**...and only 61% of gross production is marketed**

The growth in gas demand and the expansion of the gas transportation infrastructure throughout the continent (see *Chapter 3*) have resulted in a larger portion of the gas extracted being marketed: 61% in 2001, compared with 30% in 1970.

South America's total marketed production[5] of natural gas reached 107 bcm in 2001 from 63 bcm in 1991. This is equivalent to an average annual growth rate of 5.4%, well above the world's average of 2.1%. Argentina accounted for nearly half of this increase, owing to its growing domestic market and the build-up of exports to Chile. Trinidad & Tobago and Brazil experienced the fastest growth rate (9.6% and 8.8% per year respectively) though they have much lower marketed production in absolute terms.

**Table 1.3** Marketed natural gas production by country, 2001

| | Marketed production | | Aver. annual growth | |
|---|---|---|---|---|
| | bcm | % | 1981-1991 | 1991-2001 |
| Argentina | 41.5 | 39% | 8.3% | 6.7% |
| Bolivia | 4.5 | 4% | 1.8% | 2.9% |
| Brazil | 7.7 | 7% | 13.2% | 9.6% |
| Chile | 1.7 | 2% | 6.3% | 1.0% |
| Colombia | 8.0 | 7% | 2.4% | 5.9% |
| Ecuador | 0.3 | 0% | 16.0% | 0.2% |
| Peru | 0.9 | 1% | -3.3% | 2.9% |
| Trinidad & Tobago | 13.1 | 12% | 5.9% | 8.8% |
| Venezuela | 29.3 | 27% | 4.5% | 2.8% |
| **South America** | **107.0** | **100%** | **5.6%** | **5.4%** |

*Source: IEA.*

# THE NORTH-SOUTH DICHOTOMY

**The Southern Cone**

Although most of the gas reserves are located in the north of the continent, the **Southern Cone** – a region encompassing Argentina, the southern portion of Brazil, Chile, Bolivia, Paraguay and Uruguay – is where most of the action is.

**Argentina** has the most mature gas sector in the region, with well-developed fields throughout the country connected through an extensive pipeline network to all major

---

4. Flaring – i.e. combusting – natural gas produces 56 tonnes of carbon dioxide ($CO_2$) per terajoule (TJ) of dry natural gas, assuming it is fully combusted (roughly equivalent to 1.5 tonnes of $CO_2$ per million cubic metre combusted). Venting gas, on the other hand, releases uncombusted methane into the atmosphere. Venting gas is much more harmful than flaring it because the global-warming potential of methane is approximately 21 times that of $CO_2$ (over a 100-year time frame). In addition, methane contributes to tropospheric ozone problems and harms vegetation at high concentrations.
5. Data for marketed natural gas production are from the IEA; they differ slightly from those calculated by Cedigaz.

urban centres. It was the first country in South America to implement far-reaching gas-sector reforms, in the early 1990s. Growth in gas production accelerated in the 1990s, partly as a consequence of liberalisation and privatisation of the gas industry, and partly in response to increasing opportunities for exports to Chile. Argentina's marketed gas production (42 bcm in 2001) is the largest in the region, about 40% higher than Venezuela's. With proven reserves of 764 bcm at the beginning of 2002, Argentina has the third-largest reserves in South America after Venezuela and Bolivia.

In **Bolivia**, the privatisation of the gas industry in 1996 and the prospect of rapidly increasing gas exports to neighbouring energy-hungry Brazil led to a surge in gas exploration and a seven-fold increase in proven gas reserves between 1998 and 2001. With 775 bcm of proven reserves, Bolivia has now surpassed Argentina to become the country with the largest gas reserves in the Southern Cone. Although Bolivia's gas production is still rather low (4.5 bcm in 2001), it will rise gradually as the Bolivia-Brazil pipeline reaches full capacity and as other gas markets are developed. The doubling of the existing Bolivia-to-Brazil pipeline is under discussion, as well as the construction of another major pipeline from southern Bolivia to the south of Brazil. In addition, Bolivia is seeking other markets to take advantage of its huge gas reserves: there are plans to build a pipeline to the Pacific coast, either in Chile or in Peru, where Bolivian gas would be liquefied and exported to North America.

With reserves of 220 bcm, **Brazil** is rapidly stepping up domestic production to meet growing gas demand. Gross production reached 14.6 bcm in 2001, but only 7.7 bcm were marketed. According to estimates by the US Geological Survey, Brazil has the largest undiscovered resources in South America, estimated at over 5 tcm. Increased exploration activity following the opening of the oil and gas sector to private investment in 1998 will likely result in new finds. For the time being, however, Brazil will require increasing imports from Bolivia and Argentina to supply growing demand.

**Chile**'s proven gas reserves (93 bcm) are not negligible given the size of the market, but they are located in the extreme south of the country. This is a long way from the energy markets in the centre and north of the country, which can be more cheaply supplied from Argentina's Neuquén and Northeast basins. Chile's proven reserves have been falling steadily since 1984. More than half of gross production (which was 3 bcm in 2001) is reinjected; the rest is used locally as feedstock to produce methanol.

## The Northern coastline and the Andean region

**Venezuela**'s proven gas reserves (more than 4 tcm) are the largest in the region and the seventh in the world, but they have remained largely unexploited. Though the country holds more than half of South America's gas reserves, it produces only a quarter of the region's natural gas. This is due to the fact that in Venezuela 91% of proven gas reserves are associated with oil, and all gas production currently comes from oil fields. Thus, OPEC production quotas for crude oil limit the availability of gas. Furthermore, much of the gas produced is reinjected to maintain reservoir pressure in mature oil fields and to boost oil production. In an effort to increase reserves of non-associated gas and to promote gas-only production, the Venezuelan government held its first gas licensing round in 2001, allowing private investors to explore and develop areas thought to be rich in non-associated gas. Even if the country's gas production could be considerably increased, it is unlikely that Venezuelan gas will ever flow south through pipes. Thousands of kilometres, impenetrable (and environmentally sensitive) jungles, sparse population and a total lack of infrastructure (including roads) separate Venezuelan gas resources from the southern half of the continent, where 70% of South America's 330 million people live and 80% of the region's wealth is produced. Venezuela could, however, become an important exporter of liquefied natural gas (LNG) to the Atlantic market, given the right economic conditions and policy incentives.

Situated close to the Venezuelan coast, **Trinidad & Tobago** is similarly well endowed. Following major gas discoveries in 1996-2000, proven reserves have increased to 705 bcm at the beginning of 2001, doubling from their 1996 level of 349 bcm. In January 2002, the official data for proven reserves was 558 bcm. The decrease is due to a change of methodology, not a real fall in reserves. Unlike its large neighbour, the small Caribbean island has invested heavily in the development of its gas resources, which are used to fuel gas-intensive industries (ammonia, methanol) and are exported as LNG to the US and Europe. In just a few years, Trinidad & Tobago has become the world's leading exporter of ammonia, and the first LNG exporter in Latin American. It is soon expected to become a leading exporter of methanol.

In **Peru**, most of the 255 bcm of proven natural gas reserves are concentrated in the giant undeveloped Camisea field. Unfortunately for Peru, the field is located in a remote jungle east of the Andes, 300 miles from Peru's populated coastal areas and even further from primary gas users in Brazil, Chile and Argentina. Despite initial problems, the development of Camisea is now going ahead. Gas will be transported to Lima and used by power plants and industrial customers. But it may prove difficult for Peru to realise its full gas potential. The domestic market is small, and the discovery of large reserves in neighbouring Bolivia has made it unlikely that Peruvian gas will be used to supply the Brazilian market, at least in the short-to-medium term. However, there are plans to export Peruvian gas as LNG to the North American market. Here too, Peruvian gas will compete with Bolivian gas.

With proven reserves of 198 bcm, **Colombia** has been stepping up production to supply its growing gas market. The discovery of major gas reserves in the early 1990s led the Colombian government to launch a natural gas "massification" programme, entailing the rapid development of an extensive pipeline network and the increase of natural gas use in all sectors. Gross production reached 14 bcm in 2001, three times the 1990 level, but half of it is reinjected to boost oil production. Colombia has plans to further develop its gas industry and is looking at Central America as a possible export market. An interconnection with Venezuela is also under discussion. Despite Colombia's potential and the opening of its oil and gas industry to private investment in 1996, problems of insecurity may delay or discourage new investments in exploration and production.

**Ecuador**, too, has significant proven gas reserves estimated at 115 bcm in 2002 and largely associated with oil. The 11% increase between 2001 and 2002 is due to the discovery of a large 10-bcm field of non-associated gas in the Gulf of Guayaquil. The lack of necessary infrastructure means that there is currently no gas market of any significance in Ecuador. At the moment, virtually all gas production is associated with oil. Most of it is either flared or reinjected. A small portion is used for the production of liquids (LPG).

## FUTURE PROSPECTS

**Large potential for unproven and undiscovered natural gas**

Recent major discoveries in Bolivia, Peru, Trinidad & Tobago, Argentina and Brazil, not yet incorporated in proven reserves, could nearly double proven reserves in the medium term. Estimated probable and possible reserves for the whole region amount to 2.7 tcm and 2.8 tcm respectively. In addition, the region is thought to hold substantial undiscovered gas resources. According to the latest US Geological Survey (USGS) estimates, released in the *World Petroleum Assessment 2000*, undiscovered natural gas resources in South America could be as large as 13 tcm.

**Table 1.4** Total reserves and undiscovered resources (bcm)

| | Reserves, at 1 Jan. 2002 | | | Undiscovered resources* |
|---|---|---|---|---|
| | **Proven** | **Probable** | **Possible** | |
| Argentina | 764 | 305 | 282 | 1,039 |
| Bolivia | 775 | 706 | 704 | 708 |
| Brazil | 220 | 113 | | 5,505 |
| Chile | 93 | 80 | | 181 |
| Colombia | 198 | 73 | 63 | 286 |
| Ecuador | 115 | 3 | 7 | 16 |
| Peru | 255 | 262 | 450 | 179 |
| Trinidad & Tobago | 558 | 132 | 34 | 900 |
| Venezuela | 4,163 | 1,028 | 1,256 | 2,867 |
| Other | | | | 1,843 |
| **South America** | **7,140** | **2,702** | **2,796** | **13,525** |

\* *Mean estimates.*
*Sources: Reserves: Official country information, OLADE; Resources: USGS (2000).*

**Favourable R/P ratios**

Even looking only at proven reserves, natural gas has a bright future in South America. The gas reserve-to-production (R/P) ratio for the whole of South America stood at 57 years at the beginning of 2002, close to the world average and more than double that of Western and Central Europe (*Table 1.5*). In the last 20 years, the average R/P ratio for South America has oscillated between 53 and 71 years, indicating that the rate of new discoveries has been roughly in line with the increases in production.[6]

There are, however, significant differences among countries. Argentina, which is the most mature gas producer and consumer in the region, has an R/P ratio of just 19 years. The Argentine R/P ratio has steadily declined since the early 1980s when it was about 50 years. It still compares favourably, however, with other mature producers (nine years

---

6.  Definitions and measurements of proven reserves differ among companies, and hence among countries. Some companies include only gas that is contracted for sale: an increase in the demand for gas will lead to new contracts and hence an increase in reserves. Other companies include in their proven reserves all reserves earmarked for development and expected to be sold. Another factor is that it might not be profitable for oil and gas companies to prove additional reserves: if a company has enough proven reserves to fulfil its long-term commitments, it is likely to postpone drilling until reserve additions are needed. This can explain why in many countries R/P ratios based on proven reserves are roughly constant over time despite substantial increases in production. For all these reasons, R/P ratios can be misleading when looking at time trends and comparing different countries.

**Table 1.5** Proven reserves and reserve-to-production ratios, as of 1 January 2002

| | Reserves | R/P ratio** | | Reserves | R/P ratio** |
|---|---|---|---|---|---|
| | bcm | years | | bcm | years |
| **SOUTH AMERICA** | **7,140** | **57** | **FORMER SOVIET UNION** | **55,881** | **76** |
| **Southern Cone** | **1,851** | **32** | Russia | 46,475 | 79 |
| Argentina | 764 | 19 | | | |
| Bolivia | 775 | 162 | **AFRICA** | **13,107** | **74** |
| Brazil | 220 | 19 | **Northern Africa** | **7,394** | **59** |
| Chile | 93 | 68 | **Rest of Africa** | **5,713** | **111** |
| **Andean-Caribbean** | **5,289** | **80** | Nigeria | 4,500 | 132 |
| Venezuela | 4,163 | 101 | | | |
| Trinidad & Tobago | 558 | 32 | **MIDDLE EAST** | **71,138** | **262** |
| Peru | 255 | 472 | Iran | 26,500 | 338 |
| Colombia | 198 | 30 | Qatar | 25,768 | 703 |
| Ecuador | 115 | 120 | Saudi Arabia | 6,340 | 111 |
| | | | | | |
| **NORTH AMERICA** | **7,807** | **9** | **ASIA-OCEANIA** | **15,225** | **48** |
| United States | 5,350 | 9 | **China and East Asia** | **1,675** | **50** |
| Canada | 1,660 | 8 | **Indian Subcontinent** | **1,959** | **33** |
| Mexico | 797 | 17 | **South-East Asia** | **7,956** | **45** |
| | | | **Oceania** | **3,635** | **82** |
| **EUROPE*** | **7,830** | **25** | | | |
| Norway | 3,833 | 64 | | | |
| Netherlands | 1,616 | 22 | | | |
| United Kingdom | 1,111 | 10 | **WORLD TOTAL** | **178,270** | **64** |

Notes: * Europe includes Western and Central Europe.
  ** R/P ratios are calculated by dividing proven reserves (at 1/01/2002) by gross production minus reinjection (for 2001).

Sources: Cedigaz, countries' official data.

for the United States, eight years for Canada and ten years for the United Kingdom). Brazil also has an R/P ratio of just 19 years, and with demand poised to increase rapidly, the R/P ratio is likely to fall even further unless substantial gas discoveries are made. But even assuming no large new discoveries in Argentina and in Brazil, neighbouring Bolivia, with an R/P ratio of 162 years, has more than enough reserves to supply its Southern Cone neighbours for a few decades. Overall, the Southern Cone has an R/P ratio of 32 years, while the Andean region and the Caribbean, with much larger reserves and a similar level of production, have an R/P ratio of 80 years.

**Further development of reserves will depend on demand**

The pace of development of the region's gas reserves and the rate of growth of gas production will depend on gas market developments throughout the region and on opportunities for exports of liquefied natural gas (LNG) to other regions. The last ten years have shown that the limiting factor in South America is not reserves, but markets and paying demand. Few countries in South America have both large gas reserves and a correspondingly large market potential to justify development of those reserves. For example, Bolivia, Peru and Trinidad & Tobago have large gas reserves but small and limited domestic markets, although they are looking for opportunities to expand their markets both regionally and internationally. On the other hand, Brazil and Chile, where annual gas demand is growing at two-digit percentage rates, depend mainly on imports. Venezuela and Colombia, oil producers with significant associated-gas reserves, are keen to increase their domestic gas consumption, but it is doubtful that their domestic markets alone will provide enough opportunities and incentives to spur exploration and production of non-associated gas.

While the continent boasts abundant natural gas resources, these are not generally situated within easy reach of potential markets. The Andes and the Amazon forest are formidable obstacles dividing the continent along a roughly north-south and east-west axis, respectively. Seventy percent of the region's gas reserves are situated in Venezuela and Trinidad & Tobago, at the far north of the continent, separated by several thousand kilometres of impenetrable jungles and sparsely populated areas from the southern half of the continent, where the majority of South America's population lives. Abundant gas reserves have also been found in remote areas, such as the southern tip of Argentina, the Peruvian jungle and the Brazilian Amazon region. Some of these reserves are in difficult terrain or in environmentally sensitive areas. Many other isolated and little explored areas are thought to hold substantial gas resources.

Hence, in most cases, monetisation of gas reserves and gas market development will require long-distance cross-border pipelines or LNG facilities and vessels, sometimes both, as in the case of Bolivia. These large and complex projects require mammoth investments. Governments are no longer able to meet the financial needs of the gas sector, as they need to focus scarce resources on other priority public needs. The availability of private capital will therefore be crucial for South America to take full advantage of its gas bounty.

# CHAPTER 2
# NATURAL GAS DEMAND

The 1990s saw a rapid increase in natural gas demand worldwide. Given the existence of relatively widespread reserves and the inherent advantages of natural gas over other fossil fuels in terms of easier handling, more efficient burning and lower environmental impact, it is not surprising that gas is gaining market share in the energy mix of most countries around the world.

In most regions of the world, gas demand is growing primarily to meet the needs of power generation. Technological developments have made gas-fired generation in combined-cycle plants substantially more efficient and more economical compared to conventional thermal power generation. In addition, gas-fired power plants require a lower up-front investment and hence less financing than hydro or coal-fired power plants. These factors, plus the environmental advantages of gas over other fossil fuels, mean that gas is increasingly the preferred fuel in power generation.

South America is no exception to this trend. Regional gas demand increased by 5.1% per year between 1990 and 2000, while gas-fired power generation grew by 7.1% annually over the same period, well over world average growth rates of respectively 2.3% and 5.1%. Currently, South America accounts for 5% of total world gas consumption.

The most important factors and policy objectives that are driving the expansion of natural gas markets in the South America are: abundant resources; investment lags in hydropower generation due to financial constraints; growing environmental concerns; and the desire of many governments to reinforce long-term energy supply security through diversification of their energy mix.

## ECONOMIC AND ENERGY CONTEXT

**Energy demand grows faster than GDP**

Widespread structural reforms and efforts towards macroeconomic stabilisation helped South American countries recover sustained levels of economic growth in the 1990s after a "lost decade", as the 1980s have been called. Hand-in-hand with rapid economic growth, there was also a rapid increase in regional energy demand. In most South American countries, energy demand grew and continues to grow faster than GDP, and demand for electricity is growing even faster. Even in years of economic recession, energy consumption continues its upward trend. *Figure 2.1* shows the average annual growth of GDP, primary energy demand and electricity consumption during the period 1990-2000.

South America is made up of countries at various stages of economic development; per capita income levels range from less than US$3,000 (Bolivia, Ecuador) to over US$8,000 (Argentina, Chile, Trinidad & Tobago, Uruguay). Per capita consumption of primary energy is closely correlated with per capita income, with the exception of Venezuela and Trinidad & Tobago, where large indigenous oil and gas resources have allowed comparatively more energy-intensive economies to develop.

*Table 2.1* shows the average per capita levels of GDP, primary energy use and electricity consumption for each country. It is worth noting, however, that in large countries – such as Brazil and Argentina – there are very substantial variations from state to state or province to province. Moreover, in all countries there are very large differences in economic development and level of energy use between urban and rural areas, or between coastal and remote regions.

**Figure 2.1** Economic and energy growth by country, 1990-2000

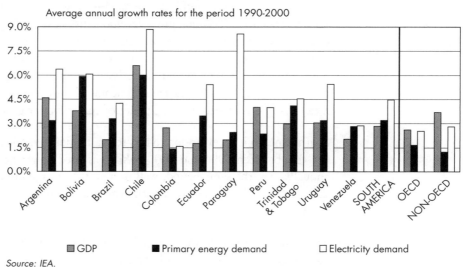

Average annual growth rates for the period 1990-2000

■ GDP    ■ Primary energy demand    □ Electricity demand

*Source: IEA.*

**The energy consumption mix is dominated by oil**

Oil still accounts for a large part of South America's energy consumption. This is not surprising since many South American countries are oil producers. In 2000, oil accounted for 45% of the region's total primary energy supply (TPES)[1], although there are significant differences among countries as shown in *Figure 2.2.* This compares with 41% for the OECD countries and 25% for Asia. The share of oil in South America's total final energy consumption (TFC) was 51% in 2000, compared with 53% for OECD countries and 28% for Asia.

Also very noticeable is South America's large use of biomass[2], which accounted for 16% of TPES and 18% of TFC in 2000. The shares of oil and biomass in TPES have both decreased over time (*Figure 2.3*), while those of hydro and natural gas increased. Indeed, natural gas has increased faster than any other fuel in both TPES and TFC in the last decade.

---

1. For definitions of total primary energy supply, total final consumption, as well as geographical groupings used in this report see the Glossary.
2. Biomass is defined as any plant matter used directly as fuel or converted into other forms before combustion. See full definition in the Glossary. In South America, biomass is mainly used in the form of wood and charcoal. Brazil also used a large amount of sugarcane to produce alcohol used in the transport sector. A limited amount of electricity is produced burning wood-waste and other vegetal wastes.

**Table 2.1** Per capita economic and energy indicators, 1990 and 2000

| | GDP[1] | | Primary Energy Consumption | | Electricity Consumption | |
|---|---|---|---|---|---|---|
| | US$ per capita | | toe per capita | | kWh per capita | |
| | **1990** | **2000** | **1990** | **2000** | **1990** | **2000** |
| Argentina | 8,376 | 11,505 | 1.38 | 1.66 | 1,308 | 2,129 |
| Bolivia | 2,000 | 2,286 | 0.42 | 0.59 | 275 | 391 |
| Brazil | 6,586 | 6,949 | 0.90 | 1.07 | 1,471 | 1,935 |
| Chile | 5,455 | 8,898 | 1.04 | 1.60 | 1,254 | 2,521 |
| Colombia | 5,408 | 5,844 | 0.72 | 0.68 | 819 | 792 |
| Ecuador | 2,982 | 2,881 | 0.57 | 0.65 | 483 | 664 |
| Paraguay | 4,413 | 4,117 | 0.73 | 0.71 | 505 | 882 |
| Peru | 3,633 | 4,518 | 0.47 | 0.49 | 553 | 687 |
| Trinidad & Tobago | 6,743 | 8,438 | 4.77 | 6.66 | 2,698 | 3,925 |
| Uruguay | 6,739 | 8,461 | 0.72 | 0.92 | 1,246 | 1,969 |
| Venezuela | 5,600 | 5,518 | 2.30 | 2.45 | 2,494 | 2,669 |
| **South America** | **6,056** | **6,821** | **0.99** | **1.15** | **1,295** | **1,708** |
| | | | | | | |
| For comparison: | | | | | | |
| *OECD Europe* | *15,791* | *18,531* | *3.26* | *3.39* | *5,021* | *5,755* |
| *OECD North America* | *21,475* | *26,321* | *6.29* | *6.70* | *9,672* | *11,183* |
| *OECD Asia-Pacific* | *18,624* | *22,014* | *3.39* | *4.30* | *5,881* | *7,932* |
| *China* | *1,668* | *3,846* | *0.77* | *0.91* | *529* | *1,016* |
| *Rest of Asia-Pacific* | *1,947* | *2,668* | *0.49* | *0.59* | *349* | *537* |

Note: 1. GDP is expressed in billion US$ at 1995 prices and power purchase parities.
Source: IEA.

**Figure 2.2** Fuel shares in total primary energy supply, 2000

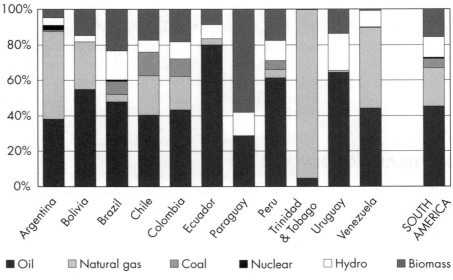

■ Oil    ▦ Natural gas    ▨ Coal    ■ Nuclear    ☐ Hydro    ▨ Biomass

Source: IEA.

**Figure 2.3** Evolution of fuel shares in TPES and TFC in South America, 1980-2000

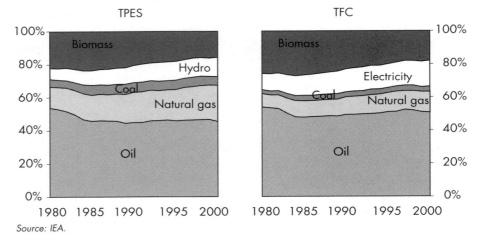

Source: IEA.

**Natural gas is the fastest-growing fuel**

Primary gas consumption in South America totalled an estimated 104.2 bcm in 2001, up 4.1% from 100.1 bcm in 2000[3]. Gas consumption increased 5.1% per year between 1990 and 2000, while total energy consumption grew 3.2% over the same period. In Brazil and Chile, which had a low level of gas use in 1990, growth of primary gas consumption averaged 12% and 14% respectively over the past decade.

As a result of this rapid growth, the share of gas in the region's primary energy mix reached 22% in 2000 from 18% in 1990, a figure comparable to the share of gas in North America and Europe (24% and 22% respectively). The regional average, however, is misleading because there are wide differences among countries, with gas penetration ranging from less than 5% (Brazil, Ecuador, Peru, Uruguay) to 45-50% in Argentina and Venezuela and 95% in Trinidad & Tobago (*Table 2.2*).

These differences are largely dependent upon the availability of domestic gas resources, but that is only part of the story. South American countries have traditionally relied on hydropower and indigenous or imported oil for their energy needs, and have had little or no interest in natural gas until recently. Even in countries with high gas potential, exploration and development efforts by state oil companies historically concentrated on oil. Until recently, little was done by governments to encourage gas-market development and expansion of the gas transport infrastructure – the only notable exception being Argentina, where the development of gas infrastructure started in the 1960s.

## NATURAL GAS DEMAND

**Drivers of natural gas demand**

Demand for natural gas is influenced by several factors, including the availability of locally-produced gas and alternative sources of energy, the cost of gas relative to alternative fuels, and environmental policies promoting the substitution of gas for other more polluting fuels.

---

3.  On a net calorific basis, this is equivalent to 100 Mtoe in 2001 and 96 Mtoe in 2000. See the Glossary for a definition of net and gross calorific value and for a discussion on the energy content of natural gas.

**Table 2.2** Primary gas consumption in South America, 1990 and 2000

| | Primary gas consumption | | | Per capita gas consumption | | Share of gas in total primary energy cons. | |
|---|---|---|---|---|---|---|---|
| | billion cubic metres | | aver. annual growth | cubic metres | | | |
| | **1990** | **2000** | | **1990** | **2000** | **1990** | **2000** |
| Argentina | 22.5 | 36.4 | 4.9% | 692 | 984 | 42% | 50% |
| Bolivia | 0.8 | 1.6 | 7.7% | 115 | 191 | 23% | 27% |
| Brazil | 2.9 | 9.0 | 12.0% | 19 | 53 | 2% | 4% |
| Chile | 1.8 | 6.5 | 13.8% | 135 | 424 | 11% | 22% |
| Colombia | 4.5 | 7.3 | 4.9% | 130 | 173 | 13% | 19% |
| Ecuador | 0.2 | 0.3 | 2.3% | 21 | 21 | 4% | 3% |
| Paraguay | – | – | – | – | – | – | – |
| Peru | 0.7 | 0.8 | 1.7% | 32 | 32 | 5% | 5% |
| Trinidad & Tobago | 5.6 | 9.8 | 5.8% | 4,620 | 7,566 | 81% | 95% |
| Uruguay | – | 0.0 | – | – | 11 | – | 1% |
| Venezuela | 21.8 | 28.4 | 2.7% | 1 115 | 1 174 | 46% | 46% |
| **South America** | **60.7** | **100.1** | **5.1%** | **206** | **289** | **18%** | **22%** |

*Source: IEA.*

Gas has some advantages and disadvantages compared with other fuels. The main disadvantage is that at normal pressure gas is about a thousand times more voluminous than oil for the same energy content. Because of this low energy density, the specific costs of gas transportation and storage are much higher than for oil. Thus, transportation is a crucial and expensive link in the gas chain; its cost can easily account for 50% of the end-user gas price. Transporting gas between point of production to the end-consumer – either by high-pressure pipelines or, after liquefaction, by ship – typically involves large investments with a high degree of inflexibility and significant economies of scale. Hence, geographical factors, such as the distance between the production site and the main consumption area, and the concentration of demand, are critical to gas market development.

On the other hand, compared with other fossil fuels, natural gas has advantages which make it attractive for certain uses despite the high transportation costs. Gas burns more cleanly than other fossil fuels, generating no ashes or sulphur emissions, and it is easier to handle, allowing much more precise and uniform combustion. Gas also contains less carbon per unit of energy output than oil or coal; thus its combustion releases less $CO_2$. Furthermore, technological advances have substantially raised the thermal efficiency[4] of gas-fired power generation compared with fuel oil and coal, making natural gas, under most circumstances, a more competitive fuel for base-load power generation.

Transport costs and climatic conditions also help explain why and where gas markets first developed. By and large, the countries with the earliest and largest use of gas

---

4.  Thermal efficiency indicates the output/input ratio of electricity produced to fuel used, with both elements measured in the same unit.

(e.g. US, FSU, Argentina, UK, continental Europe) are countries with cold temperatures in winter and where significant gas resources were found close to large urban agglomerations. The extensive use of gas in the residential, commercial and public sectors – combined with gas use in industry – provided the basis for the early development of gas transportation and distribution networks. Often in these countries, there already existed distribution networks for manufactured gas (city gas), which could be quickly converted to transport natural gas. Gas-fired generation developed later, driven mainly by environmental concerns and economics as gas was substituted for more polluting coal or displaced costly oil imports.

**Most of the gas is used in the industrial sector and the petroleum sector**

The only mature gas market in South America is Argentina, which has a well-developed infrastructure and a high level of gas penetration in all end-use sectors (see definition of gas market maturity in *Box 2.1*). Argentina has the highest residential and commercial use of gas, and a level of per capita final gas consumption comparable to Europe. In the rest of South America, there is little or no need for space heating. Hence, the use of gas elsewhere on the continent has historically been concentrated in a few oil-producing countries, where gas discovered as a result of oil exploration was used mainly in the industrial sector and in the oil and gas sector itself[5]. Gas demand in the power-generation sector has remained limited because the region has large and well-developed hydropower resources.

*Figure 2.4* shows that, for the region as a whole, there has been little change in the sectorial shares of gas demand over the last 20 years. The industrial sector remains the

**Figure 2.4** Sectorial use of natural gas in South America, 1980-2000

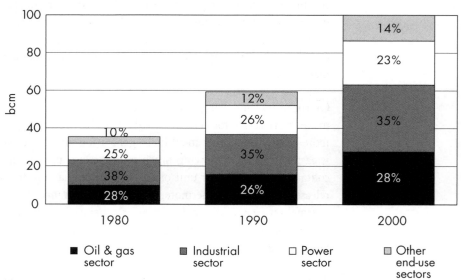

*Source: IEA.*

---

5.  For the purpose of this study, gas used in the oil and gas sector includes gas used in exploration and production, in refineries and losses in transformation and in transportation. Use of gas for methanol production is included in the industrial sector (rather than in the transformation sector as in IEA statistics). Use of gas by the oil and gas sector does not include gas that is reinjected, flared or vented.

**Box 2.1**    Gas-market maturity

The notion of market maturity for the gas sector is very different from maturity in the electricity sector. Because electricity is an essential, non-substitutable form of energy, maturity in power supply is attained with the full coverage of the country by the grid infrastructure, and with the necessary production and transport capacity in place to cover demand of the entire economy and population at all times. Natural gas differs from electricity, however, in that it is not an essential good. The heat, cooling, electric power or feedstock provided by gas can all be – and often are – obtained from other sources of energy. Economically accessible substitutes for natural gas exist in virtually all of its applications. The maturity of a gas market should thus be measured in terms of how much gas penetration can be achieved under economic conditions in competition with other sources of energy.

It is assumed that maturity is reached when gas penetration in the commercial and household sectors is advanced and can be only marginally improved, because these sectors entail the highest costs in terms of infrastructure (distribution) and supply flexibility. Maturity does not mean saturation, and there is still considerable scope for growth even in mature markets, such as Argentina or Western Europe, particularly in power generation. But the anticipated demand growth will require comparatively little investment in downstream transport infrastructure, with shorter amortisation times.

Based on this definition, Argentina and Trinidad & Tobago can be considered as having a mature downstream gas markets, as have European countries such as the Netherlands, the UK, Germany, Hungary and Italy, but also France, Poland, the Czech Republic and the Slovak Republic. Bolivia, Chile, Colombia and Venezuela still constitute "young" gas markets in the sense that much infrastructure development in both transmission and distribution will still be needed in order to reach a state of maturity comparable to the countries named above. In Europe, this is also the case for Spain, Ireland and Sweden. Brazil is a nascent gas country, in which the development of a country-wide gas supply infrastructure is still at its beginning, similar to Greece, Portugal and Turkey. Finally, in Peru, Ecuador and Paraguay, the downstream gas market is still insignificant.

Regardless of the state of market maturity, transport infrastructure is still growing in most countries, but especially in the young or nascent gas markets.
*Source: IEA, 2000.*

largest user of gas, accounting for more than a third of the region's gas supply. For the region as a whole, gas accounts for 24% of total energy use in the industrial sector, but shares vary significantly across countries (*Table 2.3*). It is interesting to note that biomass accounts for 23% and oil products for 27% of total primary energy use.

Natural gas demand in the industrial sector can be divided into three main categories: gas used as a fuel to produce direct process heat, gas used to produce steam and gas

**Table 2.3** Gas use in the industrial sector by country, 2000

| | Gas use in the industrial sector | Fuel shares in total industrial energy demand | | | | |
|---|---|---|---|---|---|---|
| | bcm | Oil | Gas | Biomass | Electricity | Coal |
| Argentina | 7.8 | 18% | 44% | 14% | 20% | 3% |
| Bolivia | 0.4 | 2% | 38% | 50% | 10% | – |
| Brazil | 5.2 | 33% | 6% | 33% | 19% | 9% |
| Chile | 3.6 | 23% | 33% | 12% | 24% | 8% |
| Colombia | 1.3 | 21% | 14% | 25% | 14% | 26% |
| Ecuador | – | 63% | – | 23% | 14% | – |
| Paraguay | – | 6% | – | 84% | 11% | – |
| Peru | – | 57% | – | 0% | 28% | 15% |
| Trinidad & Tobago | 7.1 | 1% | 95% | – | 4% | – |
| Uruguay | 0.0 | 42% | 6% | 23% | 29% | 0% |
| Venezuela | 10.0 | 20% | 61% | 2% | 15% | 1% |
| **South America** | **35.4** | **27%** | **24%** | **23%** | **18%** | **7%** |

*Source: IEA.*

used as feedstock. The industrial sector also uses gas for power generation, often in co-generation plants, i.e. plants producing both electricity and heat. However, in IEA statistics and in the graphs of this chapter, gas used for industrial power generation or co-generation is classified under power generation.

The competitiveness of gas compared with other fuels differs in the three categories. Where gas is used directly, its value is higher than that of other fuels, especially when coal and oil products cannot be used because of ash content or similar quality limitations. Even when direct heating with coal and oil products is allowed, natural gas can often claim an advantage in improved and constant product quality. For these reasons, natural gas is often the preferred fuel in the glass, ceramic, and food and beverage industries.

In steam raising, natural gas usually competes with lower-priced fuels. This is the most price-sensitive share of industrial demand. Fuel oil and coal are the other main energy sources in this market, and gas must be competitively priced if it is to gain a share of the market.

In the petrochemical and iron-and-steel industries, natural gas is used both as an energy source and as feedstock. In the petrochemical industry, methane is used in the production of methanol, ammonia and hydrogen. Methanol is the basis for producing MTBE, a high-value-added hydrocarbon, while ammonia is used in the production of fertilisers. In the iron-and-steel industry, natural gas is not only a source of heat, but may also act as a chemical reductor, thus substituting for coking coal.

In South America, the chemical and petrochemical sectors (including methanol production) accounted for about half of industrial gas demand in 2000, up from just 25% in 1980. The iron-and-steel industry accounted for about 20% of industrial gas demand.

The oil and gas sector is the region's second-largest user of natural gas, accounting for 28% of regional primary gas demand in 2000. As shown in *Figure 2.5.*, the share of the petroleum sector is largest in countries with substantial associated-gas production, such as Venezuela (42% of total primary gas supply) and Colombia (36%). In Peru and Ecuador, the small amount of gas produced is used entirely by the petroleum industry.

**Figure 2.5** Sectorial use of natural gas in South America, 2000

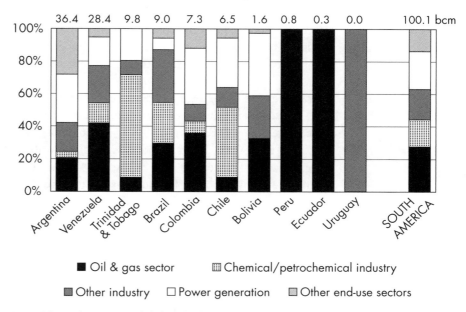

Note: Other end-use sectors include households, commerce and public services and road transport.
Source: IEA.

**The potential for gas use in the residential and commercial sectors is limited**

With the exception of Argentina, the use of gas in the region's residential, commercial and public building sectors is limited, largely because there is no need for space heating in most of the continent. In 2000, gas demand from the residential and commercial sectors accounted for 11% of total regional gas demand. Argentina and Colombia have the highest residential and commercial gas use: 24% and 11% of total primary gas supply, respectively. Further gas penetration in the residential and commercial sector is likely in Chile, where space heating is needed in winter and gas could displace imported oil and firewood.

**The use of gas as a transport fuel is growing**

The use of compressed natural gas (CNG) as a fuel for road transport has great potential in South America. Argentina already has more than 700,000 CNG vehicles. In 2000, CNG vehicles used 1.7 bcm of gas (1.4 Mtoe on a net basis), accounting for 11% of total energy use in road transport and 5% of total primary gas supply. This is the highest use of CNG in the world, in both absolute and relative terms.[6] In Brazil, the

6. By comparison, all OECD countries together used half that amount of gas (0.8 bcm, 0.7 Mtoe) in road transport in 2000.

number of CNG vehicles and service stations distributing CNG is growing very rapidly. Total CNG consumption is still low in absolute terms, but is doubling every year. In 2000, CNG accounted for 3% of total primary gas supply. At the end of 2001, Brazil had 285,000 CNG vehicles.

## Future gas demand growth will be driven by electricity generation

Over the last decade, gas-fired power generation in South America increase by 7.1% per year, well above the world average growth rate of 5.1%, and similar to the OECD average of 7.0%. Gas used in power generation increased more slowly (4.0% per year) because of the increasing efficiency of gas-fired power plants built in South America. Nevertheless, gas use for power generation is still relatively low at the regional level and concentrated in just a few countries. This is because the region has large and well-developed hydropower resources. In 2000, hydropower plants generated 75% of the region's electricity, a higher share than in any other region of the world.

Several countries in the region depend heavily on hydropower (*Figure 2.6* and *Table 2.4*). The share of hydropower in total generating capacity ranges from 30-40% in Chile, Argentina and Bolivia; to 50-70% in Colombia, Ecuador, Peru, Uruguay and Venezuela; to over 80% in Brazil and 99% in Paraguay. Recent droughts in Argentina, Chile, Brazil and Venezuela have exposed the vulnerability that comes with high dependence on hydropower. Blackouts associated with low hydropower output also occurred in Argentina in the 1980s and in Colombia in the early 1990s.

In 2000, the region consumed 23.4 bcm of natural gas (equivalent to 20 Mtoe on a net calorific basis) to generate electricity. This accounted for 23% of total gas supply in the region. The countries that use the largest share of their gas supply for power generation are Bolivia (38%), Colombia (35%), Argentina (30%) and Chile (30%). This share is the lowest in Brazil (7%). There is no gas-fired power generation in Peru, Ecuador and Uruguay at the moment. While the share of power generation in

**Figure 2.6** Power generation capacity by type of plant, 2000

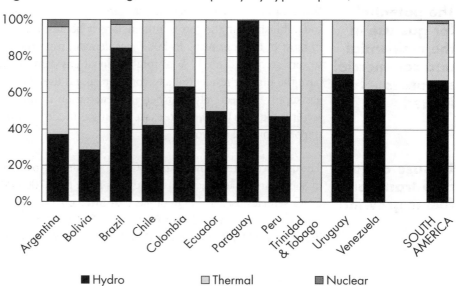

Source: OLADE, Energy Report of Latin America and the Caribbean, 2001.

**Table 2.4** Power generation mix in South America, 2000

| | Hydro | Nuclear | Gas | Oil | Coal | Other renewables* |
|---|---|---|---|---|---|---|
| Argentina | 32% | 7% | 55% | 3% | 2% | 0% |
| Bolivia | 50% | – | 46% | 3% | – | 2% |
| Brazil | 87% | 2% | 1% | 5% | 3% | 3% |
| Chile | 46% | – | 22% | 3% | 27% | 2% |
| Colombia | 73% | – | 19% | 0% | 7% | 1% |
| Ecuador | 72% | – | – | 28% | – | – |
| Paraguay | 100% | – | – | 0% | – | 0% |
| Peru | 81% | – | 4% | 13% | 1% | 1% |
| Trinidad & Tobago | – | – | 100% | – | – | 0% |
| Uruguay | 93% | – | – | 7% | – | 0% |
| Venezuela | 74% | – | 16% | 10% | – | – |
| **South America** | **75%** | **2%** | **13%** | **5%** | **4%** | **2%** |
| | | | | | | |
| For comparison: | | | | | | |
| *OECD Europe* | *17%* | *29%* | *16%* | *6%* | *30%* | *3%* |
| *OECD North America* | *13%* | *18%* | *15%* | *5%* | *47%* | *2%* |
| *OECD Asia-Pacific* | *8%* | *27%* | *19%* | *11%* | *33%* | *2%* |
| *Non-OECD Asia* | *15%* | *3%* | *10%* | *8%* | *64%* | *1%* |
| *World* | *17%* | *17%* | *17%* | *8%* | *39%* | *2%* |

Note: * Other renewables is essentially biomass in South America, but also include geothermal, solar and wind power in other regions.
Source: IEA.

total gas supply has remained unchanged over the past 20 years, gas for power is expected to be the most dynamic component of future gas demand growth, as many South American countries are now striving to reduce their dependence on hydropower.

**Brazil** has a power-generation capacity of 75 GW, of which 82% is hydropower plant (2001). Several years of below-average rains have gradually reduced the output of many hydroelectric plants, especially in the Southeast, the economic heartland of the country. The economic recession of 1998-99 eased pressure on electricity supply for a while, but the later acceleration of economic growth (4.2% in 2000) and a chronic lack of new investments in power generation and transmission brought the country to a full-blown electricity crisis in 2001. Brazil has long wanted to reduce its overwhelming dependence on hydroelectric power. In 1999, the government unveiled plans to increase the share of gas in primary energy supply from 3% to 12% by 2010, mostly as a result of an increase in gas-fired power-generation plant. According to official projections, 46 GW of new capacity will be needed between 2000 and 2010, of which 11-12 GW are likely to be gas-fired.

In **Chile**, between 1991 and 1996, hydropower plants supplied 60% to 80% of the country's electricity. However, the severe drought that gripped Chile from late 1997 well into 1999 exposed the vulnerability of the country's electricity sector. Santiago de Chile suffered repeated blackouts from November 1998 until May 1999. Chile is thus keen to diversify its electricity generation sources. Seven new gas pipelines came on stream between 1996 and 1999, bringing Argentine gas to the North, Centre and Southern regions of the country and allowing new gas-fired plants to be built. Thermal power capacity increased from 40% in the early 1990s to nearly 60% in 2000. In 2000, natural gas accounted for 22% of total power generation, up from 1% in 1996. According to the 1999 indicative plan of the Chilean Energy Commission, 83% of the 4.8 GW which are planned to be added by 2008 will be gas-fired power stations.

**Argentina** and **Bolivia** already generate about half of their electricity from natural gas. In Argentina, the share of gas in power generation reached 55% in 2000, up from 23% in 1971. Most new capacity is expected to be gas-fired. In Bolivia, the share of gas in power generation grew from less than 2% in the early 1970s to around 50% in 1996-2000. Although the country has substantial untapped hydropower capacity, there are plans to build several gas-fired plants close to the border with Brazil for export to that giant neighbour.

**Venezuela**, too, has a large hydropower capacity, accounting for 70% to 75% of annual electricity production. While there is overall over-capacity in power generation, there are substantial regional imbalances in electricity supply and demand. Furthermore, transmission constraints and network breakdowns have caused frequent localised blackouts and brownouts in recent years. This has pushed industrial users to increase their auto-generation capacity, which is often fuelled by natural gas. Whether gas-fired capacity will increase in Venezuela (either for industrial co-generation or to replace obsolete and polluting oil-fired plants) will depend on the development of the country's non-associated gas reserves, which could provide growing and predictable gas supplies. Currently, all gas production is associated with oil, and thus is constrained by OPEC quotas.

In **Colombia**, hydropower accounts for 63% of installed capacity and around 75% of power generation, but gas-fired power generation is growing rapidly as a result of the government's policies to actively promote natural gas. Gas-fired power generation capacity increased from 1.1 GW in 1994 to 3.4 GW in 2001.

**Paraguay**, **Uruguay**, **Ecuador** and **Peru** currently have little or no gas-fired capacity. Paraguay is unlikely to build any gas-fired power plants since it has enough hydropower to supply its domestic market and export a substantial surplus to Brazil. The other three countries, which all have a generating mix of about two-thirds hydro and one-third oil, will probably build some gas-fired power stations in the near future, either to use local gas resources (Ecuador, Peru) or to reduce dependence on imported oil (Uruguay).

**Trinidad & Tobago** is the only country in the region that relies on natural gas for virtually all its power generation.

While there is still considerable potential for hydroelectric expansion in several countries, the fashion has turned against large hydroelectric projects. Public budgets can no longer afford to finance large infrastructure projects, and the World Bank and other financing institutions have backed away from them, at least partly because of the environmental costs and increasing local opposition.

The institutional background has also changed. In many countries the power sector has been deregulated and opened to private investment (see *Chapter 4*). Competitive power markets do not favour capital-intensive technologies like hydropower. In most cases, private investors prefer gas-fired power stations to, for example, coal-fired or hydropower plants, because they have lower up-front costs and shorter lead times, few or no economies of scale, and can be built in small incremental modules to follow electricity demand growth.

Furthermore, many countries see gas-fired power plants as the necessary first step to develop their gas markets, as power plants provide the necessary anchor demand to spur investments in production and infrastructure. However, in hydro-dominated power systems, the competition from hydropower plants, in large part amortised, may harm the profitability of gas-fired power plants, thereby affecting the viability of the whole gas chain. Brazil is the most difficult case, as hydropower accounts for nearly 90% of electricity production and there is still substantial hydro potential. Under such conditions, it is not clear how gas-based power will compete with cheaper hydropower from already-amortised plants (and even from new hydropower plants). Where large-scale combined-cycle power stations are not competitive, other options should be explored, such as building small and medium-scale gas-fired units closer to demand, developing gas-fired distributed generation and co-generation, and encouraging the use of gas in selected electricity applications.

## FUTURE TRENDS

Gas demand grew at an average of 5.1% between 1990 and 2000. Growth accelerated in the second half of the 1990s, averaging 6.3% between 1994 and 2000 for the whole region. This high-growth trend is expected to continue in the next two decades, albeit at a slightly slower pace. The IEA's latest global demand projections estimate that gas demand in Latin America will reach 144 Mtoe in 2010 and 215 Mtoe in 2020, up from 90 Mtoe in 2000.[7] This is equivalent to annual average growth of 4.4% over the 20-year period, a much faster increase than total energy demand, which is projected to grow at 2.7% per year over the same period. As a result, the share of gas in total primary energy demand is projected to increase from 19% to 27%.

Power generation is expected to be the dominant driver for gas demand growth throughout the region. Gas-generated electricity is projected to grow at an average annual rate of 9.6% between 2000 and 2010 and 6.3% in the following decade. As a result, gas use in power generation is projected to grow from 23 Mtoe in 2000 to 86 Mtoe in 2020, rapidly surpassing gas use in industry. The share of gas-generated power in total generation is projected to grow from 12% in 2000 to 29% in 2020.

---

7. In the *World Energy Outlook 2002*, Latin America is defined as including all South and Central American countries and the Caribbean (but not Mexico, which is included in North America). This aggregate therefore includes countries which are not comprised in the aggregate called "South America" used in the rest of this chapter and throughout the report (see Glossary for definitions of geographical areas). Hence, the data for 2000 shown in this section may differ slightly from those in the rest of the chapter.

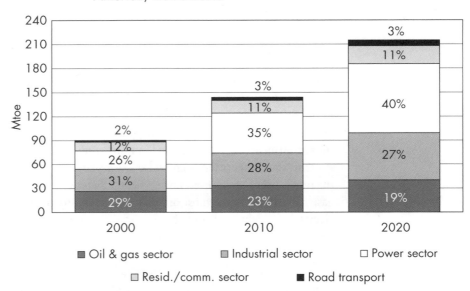

**Figure 2.7** IEA projections for natural gas sectorial demand in Latin America, 2000-2020

*Source: Adapted from IEA, World Energy Outlook 2002.*

These projections are based on a projected electricity demand growth of 3.5% between 2000 and 2010 and 3.2% between 2010 and 2020. The main uncertainty concerns the penetration of gas in the Brazilian electricity sector, for the reasons expressed above.

Gas use in the industrial sector is expected to grow from 28 Mtoe in 2000 to 59 Mtoe in 2020, a 3.8% annual growth rate. There is still significant potential for gas penetration in industry throughout the region. Several gas-rich countries still have little or no industrial gas use (Brazil, Colombia, Peru and Ecuador). In other countries, the expansion of gas networks and the implementation of environmental legislation may motivate further fuel switching by industrial consumers.

Use of compressed natural gas (CNG) in road transport is expected to increase substantially in Argentina, Brazil, and Colombia. The IEA forecasts that CNG use will grow at an annual rate of 6.6% between 2000 and 2020, rising from 2 to 7 Mtoe.

Despite plans by several South American countries to develop gas distribution infrastructure and significantly increase the number of residential and commercial gas customers over the next few years, use of natural gas in the residential, commercial and public service sectors will remain low by international standards. Except for the central and southern regions of Argentina and Chile, there is virtually no need for space heating in the region. Still, new gas-fuelled applications for space cooling and refrigeration could displace some electricity use especially in the commercial sector. Indeed, the share of gas in those sectors' energy consumption is expected to increase slightly from 11% to 15%.

**Table 2.5** IEA projections for natural gas demand in Latin America. 2000-2020

| | Energy Demand (Mtoe) | | | | Growth Rates (% per annum) | | | |
|---|---|---|---|---|---|---|---|---|
| | 1971 | 2000 | 2010 | 2020 | 1971-2000 | 2000-2010 | 2010-2020 | 2000-2020 |
| **Primary Energy** | | | | | | | | |
| **Supply** | **202** | **464** | **605** | **789** | **2.9** | **2.7** | **2.7** | **2.7** |
| of which gas | 19 | 90 | 144 | 215 | 5.6 | 4.7 | 4.1 | 4.4 |
| % of gas | 9% | 19% | 24% | 27% | | | | |
| **Final Consumption** | **155** | **361** | **465** | **602** | **3.0** | **2.6** | **2.6** | **2.6** |
| of which gas | 6 | 41 | 61 | 89 | 6.9 | 4.0 | 3.9 | 3.9 |
| % of gas | 4% | 11% | 13% | 15% | | | | |
| **Industry** | **48** | **139** | **178** | **229** | **3.7** | **2.5** | **2.6** | **2.5** |
| of which gas | 4 | 28 | 41 | 59 | 6.9 | 3.8 | 3.8 | 3.8 |
| % of gas | 8% | 20% | 23% | 26% | | | | |
| **Transportation** | **41** | **110** | **152** | **208** | **3.5** | **3.4** | **3.1** | **3.2** |
| of which gas | 0 | 2 | 4 | 7 | 7.1 | 7.1 | 6.0 | 6.6 |
| % of gas | 1% | 2% | 3% | 3% | | | | |
| **Other Sectors** | **63** | **103** | **124** | **151** | **1.7** | **1.9** | **2.0** | **2.0** |
| of which gas | 2 | 11 | 16 | 23 | 6.4 | 3.9 | 3.5 | 3.7 |
| % of gas | 3% | 11% | 13% | 15% | | | | |

| | Electricity Generation (TWh) | | | | Growth Rates (% per annum) | | | |
|---|---|---|---|---|---|---|---|---|
| | 1971 | 2000 | 2010 | 2020 | 1971-2000 | 2000-2010 | 2010-2020 | 2000-2020 |
| **Total Generation** | **135** | **804** | **1135** | **1566** | **6.4** | **3.5** | **3.3** | **3.4** |
| of which gas | 14 | 98 | 247 | 454 | 6.9 | 9.6 | 6.3 | 7.9 |
| % of gas | 11% | 12% | 22% | 29% | | | | |

| | Power Gen. Capacity (GW) | | | Growth Rates (% per annum) | | |
|---|---|---|---|---|---|---|
| | 1999 | 2010 | 2020 | 1999-2010 | 2010-2020 | 1999-2020 |
| **Total Capacity** | **168** | **256** | **357** | **3.9** | **3.4** | **3.8** |
| of which gas | 22 | 68 | 124 | 12.0 | 6.3 | 9.1 |
| % of gas | 13% | 26% | 35% | | | |

*Source: Adapted from IEA, World Energy Outlook 2002.*

Clearly, future gas demand prospects vary widely across countries, as can be seen in *Figure 2.8*, which summarises the results of a study by the Latin American Energy Organisation (OLADE) published in January 2001. In **Argentina** and **Venezuela**, where gas already accounts for a high share of energy demand, further growth is expected to be moderate. Projections for Argentina made before the 2001 economic and financial crisis showed an average annual growth of 5.6% between 2000 and 2010; this figure will have to be revised downwards to take into account the effects of the crisis on energy

**Figure 2.8** Expected growth in gas demand, 2000-2015

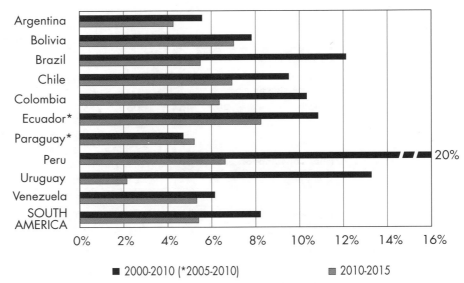

Source: OLADE, Study for Natural Gas Market Integration in South America, 2001.

supply and demand. The key engines of growth will be **Brazil, Colombia** and **Chile,** with expected demand growth of 12%, 10% and 9.5% per year respectively between 2000 and 2010. In all three countries, power generation will be the main driver for gas demand growth. Together, these three countries are expected to account for 50% of the region's incremental gas demand between 2000 and 2010. High growth rates in Ecuador, Paraguay, Peru and Uruguay are due to their very low or nil level of initial gas consumption; absolute levels remain small.

# CHAPTER 3
# NATURAL GAS TRADE

## THE SLOW PATH TO ENERGY INTEGRATION

Historically, there has been very little intra-regional energy trade in South America. This is due in part to the large distances involved and to the phenomenal geographical obstacles: the continent is divided by the Amazon forest along a roughly east-west line, and by the Andes along a north-south line.

The main reason, however, for the historical absence of significant energy exchanges and interconnections is the relatively good energy endowment of individual countries. Nearly all countries in South America have hydropower resources and at least some hydrocarbon resources. All have significant biomass resources and considerable potential for other renewables, such as solar and wind. Until recently, regional and national political conditions were such that countries tended to rely on their own energy resources.

Some cross-border electricity interconnections existed, but they were usually of limited capacity and built to provide backup to remote or isolated border systems, rather than to optimise the use of combined resources. Intra-regional trade in crude oil and refined products by pipeline or trucks was also limited.

Stronger energy linkages and cooperation emerged at the end of the 1970s in the Southern Cone, with the construction of three large binational hydroelectric power stations:

■ Salto Grande, on the Uruguay River on the Argentina-Uruguay border, 1.9 GW, started operation in 1979;

■ Itaipú, on the Paraná River on the Brazil-Paraguay border, 12.6 GW, stated up in 1985;

■ Yaciretá, further downstream on the Paraná River on the Argentina-Paraguay border, 3 GW, was commissioned in 1994.

As can be seen in *Figure 3.1*, these binational hydro projects greatly increased electricity trade in the Southern Cone. Whilst ownership of the plants is shared equally by the neighbouring countries, Uruguay and Paraguay use very little of the plants' output, and thus "export" most of their share of electricity production to their large neighbours.

Natural gas trade has taken much longer to emerge. This is because, throughout the region, interest in natural gas *per se* is relatively recent. Even in countries with large

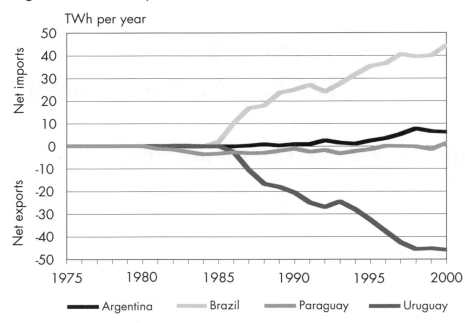

**Figure 3.1** Electricity trade in the Southern Cone, 1975-2000

*Source: IEA.*

hydrocarbon resources, production and commercialisation of natural gas was not a priority for the state-owned national oil companies, which preferred to concentrate on the development of oil resources.

Also, natural gas transportation involves much higher costs and complexities than oil transportation (*Box 3.1* and *Figure 3.2*). In South America, long distances and difficult terrain generally separate known gas reserves from main population and industrial centres. When gas reserves and gas markets are in two different countries, the challenges are even greater. All these factors contributed to discourage gas projects in a region relatively well-endowed with other energy sources.

For 30 years, exports of Bolivian gas to Argentina remained the sole case of gas trade in South America. In the 1960s, when the Argentine gas market was expanding rapidly, it seemed that domestic reserves would not be sufficient to keep up with demand growth. On the other side of the border, Bolivia had substantial associated-gas reserves with no market outlet. The pipeline linking Argentina and Bolivia was the first cross-border gas project in South America and, until the mid-1990s, the only one. Bolivian exports to Argentina, however, never exceeded 2 bcm per year – the amount agreed in the initial 20-year contract – because shortly after the start of exports, Argentina found abundant gas reserves on its own territory. Nevertheless, gas exports provided Bolivia significant income and the drive to develop further its oil and gas industry.

A more favourable environment for energy integration – and for cross-border gas interconnections in particular – emerged in the 1990s, due to a more stable political and economic environment, growing economies and dramatic economic and energy

**Box 3.1**   The specificities of natural gas transportation

Natural gas at normal pressure is about a thousand times more voluminous than oil for the same energy content. Even under high pressure (e.g. 100 bar), the energy density of natural gas is about one-tenth that of oil: this means that an oil pipeline can transport ten times as much energy as a gas pipeline of the same diameter. Thus, transportation makes up a greater share of the final price of gas than is the case for other fuels.

Gas is generally transported in one of two ways:

- Via a pressurised pipeline from the producing well to a gas processing plant (where impurities and liquids are extracted); then by a high-pressure transmission pipeline and, in case of small consumers, by a lower-pressure distribution pipeline to the end consumer; or

- Following transportation to a liquefaction plant, the gas is treated and cooled to around -160° until it becomes a liquid. This reduces its volume by a factor of around 600. The liquid natural gas (LNG) is then transported by special vessels to a receiving terminal, where it is regasified and then distributed to end-users by means of pipelines.

The relative economics of these two methods of transportation depend on the volume of gas to be transported and on the distance and type of terrain to be crossed. *Figure 3.2* illustrates the costs of transportation by pipeline and as LNG, in comparison with oil transportation by pipeline and by tanker. For a given amount of gas throughput, there is a point at which LNG transportation becomes cheaper than pipeline transportation. The distance at which this point is reached depends upon the terrain where the pipeline is to be laid, which influences the cost of the pipeline. Onshore pipelines over relatively flat terrain are at the cheaper end of the spectrum, while offshore pipelines in deep water are the most expensive. Whether gas in transported by pipeline or as LNG, the specific transportation costs are more than ten times higher than for oil transportation.

**Figure 3.2** Representative costs of oil and gas transportation

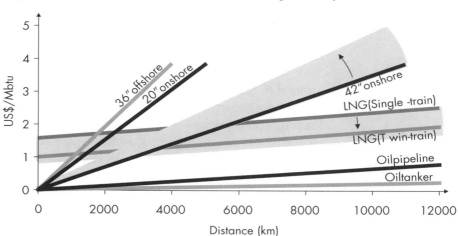

*Sources: Pipeline costs: Jensen (2000); LNG costs: IEA.*

reforms, leading in most cases to the privatisation of state-owned companies and the development of new regulatory and fiscal frameworks (see *Chapter 4*).

As in other parts of the world, technological developments and environmental concerns contributed to a growing interest in natural gas. Private-sector players, often regional or international companies, were quick to identify and take advantage of the business opportunities offered by the complementaries between gas-rich countries with small or no markets and large potential markets with no domestic gas supply. In just a few years, natural gas has emerged as the driving force for regional energy integration, particularly in the Southern Cone.

## GAS EXPORTERS AND IMPORTERS

The previous two chapters pointed out that few South American countries have both large reserves and a correspondingly large potential domestic market necessary to justify the large costs involved in developing those reserves and in building the transportation infrastructure. With respect to gas reserves and market potential, South American countries broadly fall into one of the four categories identified in *Figure 3.3*.

**Figure 3.3** Gas reserves and potential of domestic markets

| | | Domestic gas market potential | |
| --- | --- | --- | --- |
| | | **High** | **Low** |
| **Level of domestic gas reserves** | **High** | Argentina, Venezuela, Colombia | Bolivia, Trinidad & Tobago, Peru, Ecuador |
| | **Low** | Brazil, Chile | Paraguay, Uruguay |

Bolivia, Ecuador, Peru, Trinidad & Tobago have large reserves but small and limited potential domestic gas markets. To take full advantage of their gas resources, they must develop export markets. The location of their gas resources (whether with easy sea access or close to a large demand centre in a neighbouring country) will determine whether exports will be via pipeline or as LNG.

Trinidad & Tobago and Bolivia have made the most progress in developing their gas resources for export. In just a few years, Trinidad & Tobago has become the first South American country – and the only one so far – to export LNG, and is a leading exporter of gas-intensive products, such as ammonia and methanol. Bolivia was the first exporter of natural gas in the region, through a pipeline connection to Argentina built in the early 1970s. Bolivia currently exports piped gas to Brazil and is considering LNG exports to the Pacific market, by piping the gas to a port in Chile or Peru where a liquefaction plant would be built. A pipeline to Chile would also allow gas exports to Chile. In the longer term, Bolivia may resume gas exports to Argentina, if Argentine reserves prove insufficient.

In Brazil and Chile, annual gas demand is growing at two-digit percentage rates. They do not have sufficient or conveniently-located reserves to supply their growing internal demand. They will therefore rely increasingly on piped imports from neighbouring gas-rich Bolivia and Argentina. Brazil may also import some LNG. In the longer term, however, if Brazilian undiscovered resources are transformed into reserves, Brazil may move to the "high-reserve/high-market-potential" category.

Argentina, Venezuela and, to a lesser extent, Colombia have large reserves and high domestic market potential, but they differ greatly from one another. Argentina has substantial reserves and a highly developed gas market with large, and still rising, demand. It exports increasing volumes to Chile and has just started exporting gas to Brazil. Its reserve-to-production ratio of less than 20 years will decrease quickly as exports to Chile and Brazil build up. Some analysts argue that if new gas reserves are not found and developed in the near future, Argentina may become a net gas importer before long.

Venezuela has large associated-gas reserves and uses most of the associated-gas production for reinjection and as a fuel in the oil and gas sector. The government is keen to develop non-associated gas reserves so as to decouple gas production from oil production. While there is significant potential to expand domestic gas consumption, it is unlikely that the domestic market alone (given the perceived economic and political risks) will provide enough incentives to spur exploration and production of non-associated gas.

Given its geographical location, Venezuela has the potential to become a major LNG exporter to the Atlantic market, and especially to the United States. But timing is crucial, as Venezuela will face stiff competition not only from neighbouring Trinidad & Tobago, whose LNG infrastructure is already in place and expanding rapidly, but also from other existing and projected LNG sources all aiming at supplying the US East Coast. Venezuela has long been entertaining the idea of a cross-Caribbean pipeline, but the economics seem to favour transporting LNG by barge.

Colombia has medium-size gas reserves (mostly associated with oil, but some important non-associated gas fields have been found recently) and good market potential. Depending on the pace of development of gas reserves and the rate of growth of gas demand, Colombia could become either a gas exporter (to Central America, for example) or a gas importer (from Venezuela). In the medium term, however, it is likely that Colombia will actually export gas to Venezuela, which needs additional gas in the Maracaibo area, mainly for reinjection into its oil wells.

Both Colombia and Venezuela are eyeing the Central American market, which has no gas supply at the moment. Colombia is better placed, as it borders on Panama; Venezuelan gas would have to go through Colombia. In this respect, Venezuela's proposal for a cross-border pipeline to import gas from Colombia could in reality secure a route for future Venezuelan exports to Central America.

## CROSS-BORDER PIPELINE TRADE

**Rapidly-growing pipeline gas trade in the Southern Cone**

Gas pipeline trade is increasing rapidly in the Southern Cone, a region encompassing the southern half of Brazil, Argentina, Chile, Bolivia, Paraguay and Uruguay. This is where most of the population and industrial infrastructure is located and where the growth in energy and gas demand is highest. As mentioned in the previous section, both Argentina and Bolivia have abundant non-associated gas reserves, which they are eager to export to neighbouring countries, while Brazil and Chile have high-growth markets.

*Figure 3.4* illustrates the rapid growth of cross-border gas trade in this area, from 2.1 bcm in 1996 to 9.9 bcm in 2001. In that year, cross-border flows accounted for 18% of the Southern Cone's marketed gas production. Much of this was accounted for by Argentina's exports to Chile (5.3 bcm in 2001); additionally, 3.7 bcm went from Bolivia to Brazil, 0.74 bcm from Argentina to Brazil, 0.12 bcm from Bolivia to Argentina and 0.04 bcm from Argentina to Uruguay.

Between 1996 and 2001, seven pipelines were built between Argentina and Chile, the 3,150-km Bolivia-to-Brazil pipeline was finalised, and the first stage of an Argentina-to-Brazil pipeline became operational, creating the basis for a sub-regional gas transportation network. Another pipeline from Argentina to southern Brazil via Uruguay is at an advanced planning stage and a possible route from southern Bolivia to Brazil via Argentina and Paraguay is being considered.

*Map 3*[1] gives an overview of existing pipelines, as well as those currently under construction or at an advanced stage of planning.

**Figure 3.4** Cross-border gas trade in the Southern Cone, 1991-2001

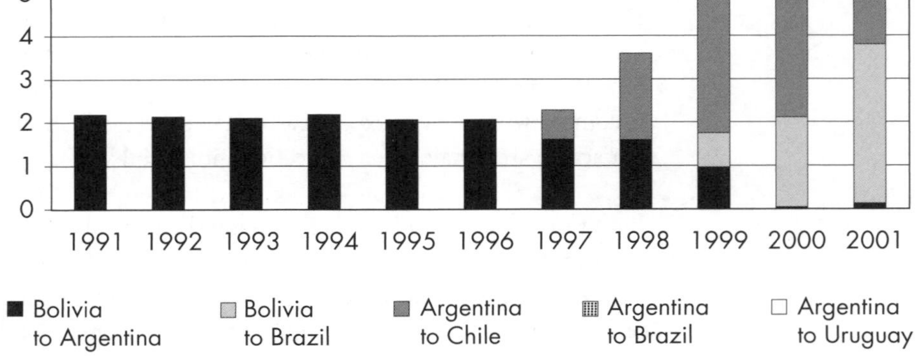

gas exports in bcm per year

■ Bolivia to Argentina  ▨ Bolivia to Brazil  ▨ Argentina to Chile  ▦ Argentina to Brazil  □ Argentina to Uruguay

*Sources: YPFB (for Bolivia), Secretaría de Energía y Minas (for Argentina).*

---

1.  All maps are located at the end of the book.

## Existing cross-border gas pipelines

### Bolivia-Argentina

Until the mid-1990s, the pipeline linking Argentina and Bolivia was the only cross-border gas pipeline in South America. The 441-km, 24-inch trunk line known as *Yabog* (Yacimientos-Bolivian Gulf) was commissioned in 1972 and has a capacity of 6 mcm/d. It connects Río Grande, in the gas-rich region of Santa Cruz de la Sierra (Bolivia), to the Argentine northern transportation network in Campo Duran (Salta, Argentina), crossing the border at Yacuiba/Pocitos. From 1972 to June 1999, when the export contract expired, Bolivia exported a total of 53 bcm, worth about US$4.5 billion.

Currently, Bolivia exports small quantities of gas to Argentina through two short cross-border pipelines owned by Argentine company Pluspetrol. These pipelines link the Pluspetrol's Bolivian fields of Bermejo y Madrejones, situated close to the border with Argentina in Tarija, to the company's facilities located on the other side of the border in Ramos and Campo Durán (Salta province).

### Argentina-Chile

Cross-border gas trade started in earnest only at the end of the 1990s, with Argentina's exports to Chile. Between 1996 and 1999, seven pipelines were built between the two countries:

- three small pipelines in the extreme south of the continent, linking Southern Argentina's Austral Basin to Punta Arenas in Chile (*Tierra del Fuego*, *El Cóndor-Posesión* and *Patagónico*);

- two pipelines supplying gas from Argentina's Neuquén Basin to Chile's central and southern regions (*GasAndes* and *Pacífico*); and

- two pipelines bringing gas to the north of Chile (*GasAtacama* and *NorAndino*).

The first cross-border pipeline connecting Argentina to Chile was the 83-km, 14-inch *Tierra del Fuego* pipeline in the far south, also called the *Methanex PA*. With a capacity of 2 mcm/d, it was built to meet the increased gas demand resulting from the expansion of a methanol plant in the extreme south of Chile. It came on stream in 1996. The other two pipelines in the far south, the 9-km *El Cóndor-Posesión* and the 33-km *Patagónico* (also called *Methanex YPF* and *Methanex SIP*), began operation in 1999 for the same purpose and have throughput capacities of 2 and 2.8 mcm/d, respectively.

The 463-km, 24-inch *GasAndes* pipeline, which extends from La Mora, Mendoza, in west-central Argentina to San Bernardo, on the outskirts of the Chilean capital, Santiago, began operating in August 1997. It cost US$325 million. With a throughput capacity of 9 mcm/d, it supplies gas from Argentina's Neuquén Basin mainly to Metrogas, the gas distribution company for Santiago, and to a 379-MW power plant in Nueva Renca, Chile.

The 638-km *Gasoducto del Pacífico* transports gas from Argentina's Neuquén area to the Bio-Bio region in southern Chile. Commercial operations began in November 1999. The cost of the pipeline was estimated at US$350 million. It has diameters of 10, 12, 20 and 24 inches and a throughput capacity of 9.7 mcm/d. The gas comes from the Neuquén Basin and supplies industrial clients (cellulose, paper, cement, steel and glass factories, and fish and sugar processing plants). It also supplies commercial and residential users in Concepción, the second-largest city in Chile.

The 941-km, 24-inch, 8.5-mcm/d *GasAtacama* pipeline connects the Argentine network in Corneja, Salta, to the city of Mejillones on the northern coast of Chile. It was completed in July 1999 and supplies gas from Argentina's gas fields in the Noroeste Basin to power plants and industrial clients in the north of Chile. In addition to the natural gas pipeline, the GasAtacama project included two gas-fired power plants in Chile.

The 1,066-km, 12-, 16- and 20-inch *NorAndino* pipeline also links Argentina's Noroeste Basin in Salta Province to Mejillones in Chile's northernmost region. With a transportation capacity of 7.1 mcm/d, it supplies industrial clients in Chile's Atacama Desert region in the north, including the Tocopilla power plant and the world's largest copper-mining industry. It began operating in November 1999.

Argentina's gas exports to Chile grew steadily from 0.67 bcm in 1997 to 5.3 bcm in 2001.

**Argentina-Uruguay**

In addition, Argentina has started to export gas to Uruguay and Brazil. The 26-km, 0.7-mcm/d *Gasoducto del Litoral*, linking the Argentine province of Entre Ríos to Paysandú in Uruguay, began operating in late 1998. A second branch was commissioned in 2000 to supply a 360-MW thermal power plant at Casablanca, Uruguay.

Construction of a new pipeline linking Buenos Aires with Colonia and Montevideo in Uruguay started in March 2001. This 208-km, 5-mcm/d *Cruz del Sur* pipeline will cost an estimated US$170 million. First gas deliveries were expected in 2002, but the project has been delayed by the economic crisis affecting both Argentina and Uruguay. The source of the gas will be the Neuquén Basin in Argentina. While the initial target of the project is the Uruguayan market, Cruz del Sur is the first segment of a bigger project to link Argentina to southern Brazil.

**Argentina-Brazil**

Since July 2000, Argentina has exported gas to Brazil through a 440-km 24-inch pipeline connecting the TGN network at Aldea Brasilera, Paraná, in the centre-east of Argentina to Uruguaiana, a Brazilian town on the border with Argentina. The *Paraná-Uruguayana* pipeline supplies a 600-MW thermoelectric power station at Uruguaiana. This is the first stage of a more ambitious project to supply Argentine gas to Porto Alegre in southern Brazil, thereby competing with Bolivian gas.

**Bolivia-Brazil**

In 1999, Bolivian gas began to flow to Brazil through what is widely considered the cornerstone of the nascent integrated gas grid in the Southern Cone. The 3,150-km *Gasbol* pipeline is the longest in South America. Built in two stages at a total cost of US$2.1 billion, it will ultimately have a capacity of 30 mcm/d.

The first stretch of the pipeline, 1,418 kilometres long with a diameter varying from 32 inches to 24 inches, started operation in June 1999. It begins at Rio Grande in Bolivia, crosses into Brazil at Puerto Soares-Corumbá in the state of Mato Grosso do Sul, reaches Campinas in the state of São Paulo and continues to Guararema, where it connects with the Brazilian network that supplies the cities of São Paulo, Rio de Janeiro and Belo Horizonte. The second stretch, linking Campinas to Canoas, near Porto Alegre in the state of Rio Grande do Sul, was completed in March 2000. It is 1,165 kilometres long with a diameter varying from 24 inches to 16 inches.

A second Bolivia-Brazil pipeline started operating in 2002. This is a 626-km spur off the Gasbol pipeline, sometimes referred to as the *Lateral Cuiabá*. It starts at Rio San Miguel in Bolivia, crosses into Brazil at Cáceres and proceeds to Cuiabá in the State of Mato Grosso. In Cuiabá, it fuels a 480-MW thermal-power station. The pipeline (360 kilometres in Bolivia, 266 kilometres in Brazil) has a diameter of 18 inches and a capacity of 2.8 mcm/d. It will transport gas from Bolivia, but also from Argentina: the Cuiabá power station already has a supply contract with an Argentine producer.

## Cross-border gas pipelines under construction or planned

Several other pipelines to transport Argentine gas to southern and southeastern Brazil's industrial and population centres have been proposed or are at various stages of development.

The most advanced is the extension of the *Paraná-Uruguayana* pipeline to Porto Alegre in southern Brazil. When concluded, the 24-inch pipeline will extend 1,055 kilometres and eventually have a capacity of 12 to 15 mcm/d. The pipeline was expected to be operational by mid-2002, but construction has been delayed for nearly two years because of uncertainties both on the demand side in Brazil and on the supply side in Argentina. In Brazil, uncertainties related to the definition of new electricity and gas sector regulations have delayed the construction of two gas-fired power stations near Porto Alegre, which will take some 5 mcm/d of the pipeline's capacity. In Argentina, a capacity expansion of the northeastern transportation network is needed to supply the new line, but the current economic crisis has halted all capacity expansion plans.

The other project to supply Argentine gas to Brazil is an extension of the *Cruz del Sur* pipeline. Starting in Colonia, Uruguay, this second spur will extend 415 kilometres in Uruguay and 505 kilometres in Brazil to Porto Alegre. The US$400 million project aims to supply 12 mcm/d to the same market targeted by the *Paraná-Uruguayana-Porto Alegre* pipeline.

It is not clear that southern Brazil's gas demand would justify two pipelines: the region has substantial hydropower and also cheap local coal. But the ultimate goal is to supply the much bigger markets of São Paulo and Rio de Janeiro, where Argentine gas could successfully compete with Bolivian gas.

Other Argentine-Brazil cross-border projects that have been discussed are the *Mercosur* and the *Trans-Iguaçu* pipelines. The *Mercosur* project envisages linking Salta in northwestern Argentina to Asunción, Paraguay, then to Curitiba, Brazil, and eventually to São Paulo. The pipeline would be 3,100 kilometres long and have a capacity of 25 mcm/d. The *Trans-Iguaçu* pipeline would cross from Argentina's Noroeste Basin into southern Brazil, with a capacity of 35 mcm/d.

While there were doubts about the adequacy of reserves in Argentina's Noroeste Basin to support such large investments in pipelines, the newly-found abundant reserves in southern Bolivia are reviving interest in a route that would link the reserves in southern Bolivia and north-western Argentina to the rich potential markets of south and southeastern Brazil. The latest project to be announced is the so-called Gas Integration (*Gasin*) pipeline, a US$5-billion undertaking that would link Bolivia, Argentina,

Paraguay and Brazil. The 5,250-km pipeline would start near the gas fields in southern Bolivia and run through gas-rich northern Argentina, where a spur would cross the border to Asunción, the capital of Paraguay. The pipeline would then enter Brazil at the western border of Santa Catarina State and run 3,450 kilometres inside Brazil, east and then north all the way to the federal capital Brasilia, supplying a number of large urban agglomerations along the way. According to initial plans, the feasibility study is to be concluded in 2002, and the pipeline could be operational by 2005.

**Only one cross-border gas pipeline project in the Andean countries**

All these projects are located in the Southern Cone. In the northern part of the continent, the scope for gas interconnections is much less, as all countries (Venezuela, Colombia, Ecuador, and Peru) have enough gas resources to supply their needs. Nevertheless, there are plans to build a cross-border pipeline between western Venezuela and Colombia. The proposed pipeline would be approximately 200 kilometres in length and would connect the offshore fields and transport infrastructure of the Guajira region in northeast Colombia to markets in Lake Maracaibo, Venezuela. The pipeline is expected to start operating in 2005, supplying Colombian gas to Venezuela for a minimum of 7 years. Venezuela, which holds the eighth-largest gas reserves in the world, faces a gas shortage because its associated-gas production is constrained by OPEC oil production quotas. The Lake Maracaibo area, where many of Venezuela's mature oil fields are concentrated, needs additional gas for reinjection, until domestic non-associated gas production can be developed. In the longer term, the flow in the pipeline could be reversed to allow Venezuelan exports to Colombia, whose gas demand is expected to quickly outgrow domestic reserves.

## LNG TRADE

**LNG exports to the Atlantic market**

While the large reserves in the north are too far away to supply pipeline gas to the Southern Cone markets, they offer great potential for LNG projects. Trinidad & Tobago inaugurated its first LNG train in 2000 and the second one started operation in 2002. In 2001, Trinidad & Tobago exported 3.8 bcm of natural gas[2], equivalent to 29% of its total marketed gas production. LNG from Trinidad & Tobago currently supplies the US East Coast, Puerto Rico and Spain. Construction is continuing on a third and fourth train, and there are plans to add a fifth train by 2005, which would make the small Caribbean island one of the largest LNG suppliers of the Atlantic market.

Venezuela certainly has enough gas reserves to become a major LNG exporter to the Atlantic market, while meeting greatly increased domestic demand. Projects for LNG liquefaction plants in Venezuela have long been stalled by poor economics and a lack of political support. Technological developments, which are lowering the costs in the LNG chain, as well as a new policy favouring gas projects to reduce the country's dependence on oil exports, are giving new impetus to LNG projects. The government has recently awarded several concessions to develop offshore gas reserves, whose output will be used partly in LNG export projects.

---

2. Cedigaz (2002).

**LNG exports to the Pacific market**

Further south, and looking at the Pacific market, Bolivia is seeking to capitalise on its enormous gas reserves and is exploring the possibility of exporting LNG to the west coast of Mexico and/or the United States via a pipeline to a port in Chile or in Peru, where a liquefaction plant would be built. Peru is also looking at the Pacific LNG market, since the local market is very small and pipeline exports to Brazil now seem precluded by the abundance of gas in Bolivia.

**LNG imports**

As yet, LNG is not imported. However, there are plans to construct an LNG importing terminal on the eastern coast of Brazil. LNG could come from Trinidad & Tobago, Nigeria or eventually Venezuela. Some analysts speculate that in the longer term LNG may also flow to Argentina and/or Chile, if gas exploration in Argentina is disappointing. This seems unlikely with so much gas recently discovered in Bolivia and Peru, and the large potential for undiscovered gas in Brazil.

## FUTURE PROSPECTS

**Regional gas demand will lead to increasing gas trade in the Southern Cone...**

*Map 4* illustrates the current natural gas trade flows in the region, while *Map 5* illustrates the likely situation in 2010. In the Southern Cone, gas pipeline trade will continue to expand, driven by the increasing gas demand of Brazil and Chile, mainly for power generation. Additional supply will come from Argentina and Bolivia. In the latter country, several new giant fields will be brought into production in the next few years.

One of the main uncertainties affecting the pace of development of the Southern Cone's pipeline infrastructure is future gas demand in the Brazilian power sector. In that country, regulatory uncertainties, gas-pricing issues and the inherent complexity of introducing gas-fired combined-cycle plants in a hydro-dominated system are delaying thermal power projects with a total capacity of several thousand megawatts, and hence delaying the construction of new gas pipeline. The other uncertainty concerns the economic and financial situation in Argentina and its effects on the investment climate not only in Argentina, but also in neighbouring countries.

Gas trade flows among the Andean countries are more uncertain because, unlike in the Southern Cone, there are no evident energy complementarities between these countries, which all have oil and gas resources of their own.

**...while gas demand from North America will drive LNG exports from the Caribbean and the Pacific coasts**

South America will likely become a significant LNG exporter in the long term. Some South American LNG will continue to flow to Europe. But the main market for South American LNG from the Caribbean (Trinidad & Tobago, Venezuela) and from the Pacific coast (Bolivia, Peru) is likely to remain North America, more particularly, the United States. Most observers agree that the United States' traditional sources of gas – domestic production and imports from Canada – will not be able to keep up with demand, which is expected to increase substantially over the next decade, especially for power generation. Some South American LNG is likely to be imported also by Mexico, either for re-export to the US or for supplying its rapidly-growing domestic market, where demand is quickly outstripping production capacity. Several projects are under way in Mexico and in the US to build new LNG-receiving terminals or reactivate terminals which had been mothballed.

The main uncertainty here is related to demand and, more importantly, to gas prices in the United States, which have proven difficult to predict in the past, and which will determine the profitability and economic viability of South American LNG projects. Other factors are uncertainty about the construction of new LNG-receiving terminals in the US and competition from other existing and projected LNG projects in Nigeria, Alaska, the Canadian Arctic and even Australia.

An alternative way to commercialise gas reserves that are far from markets in South America and without convenient access to the coast is through gas-to-liquid (GTL) schemes. The competitiveness of GTL versus LNG will depend on further technological developments to reduce costs of GTL and on the individual circumstances of each project.

# CHAPTER 4
# GAS-SECTOR POLICY AND REFORMS

## A GLOBAL MOVE TOWARDS GAS-SECTOR REFORMS

The structure and organisation of the gas industry varies significantly around the world, ranging from countries in which one company is responsible for the whole gas chain, from production to burner tip, to countries where producers compete and many companies are involved in production, transmission, distribution and supply. In some countries, public authorities (federal, provincial or municipal) own most of the gas chain; in others, it is mainly privately owned. The differences across countries regarding gas-industry structure, organisation and ownership result from many historical and geographic factors, including the extent to which a country is self-sufficient in gas or dependent on imports; the geographical distribution of gas supplies (whether there are many small fields or a few large fields) and the number of suppliers; the maturity of the gas market and the number of consumers; the availability of other energy sources; and differing national policy choices regarding liberalisation and state ownership.

In the past, direct state participation in the gas industry, through state-owned enterprises (SOE), was the norm in most countries[1]. Upstream, SOE were considered an effective instrument to manage national resources and efficiently extract and re-allocate the rent they generate. Downstream, SOE were thought to be essential to guarantee the delivery of services considered as public goods. Vertically-integrated monopolies characterised most gas markets in their early stages of development, because of the required high investment and long lead times, technical and financial risks, and the low returns associated with building new pipelines and distribution networks.

This model came under increasing pressure in the 1980s in both industrialised and developing countries, albeit for different reasons. In industrialised countries, gas-sector reforms were the consequence of a broader "philosophical" shift in economic policy, driven by globalisation, calling for a greater reliance on market forces and less direct intervention by the state. According to the new economic paradigm, commercial activities should be left as much as possible to private initiative, while the state should limit itself to defining the framework to ensure that its policy objectives are delivered, while addressing and compensating for market failures and externalities, where they exist.

---

1.  With notable exceptions like the US and Germany, whose gas sectors were and still are characterised by a large number of private companies.

Most industrialised countries have mature gas markets, with significant infrastructure already in place and modest energy demand growth. In those countries, the prime motivation for gas-sector reforms was to improve the sector's economic efficiency and lower the cost of gas for consumers by introducing competition and customer choice. Indeed, once the large capital investments in infrastructure have been amortised, operating costs and risks tend to fall and gas companies' profits tend to rise. The pressure for change came from other gas producers and consumers eager to capture a share of the economic rent. Local and regional price disparities are typically among the reasons for large users to push for liberalisation and competition. Another objective of gas reforms in industrialised countries is to transfer risks of investment to the private sector, which in developed countries is well-developed and fully capable of assuming such risks.

In developing countries, where gas markets are usually at an early stage of development and energy demand is still growing at high rates, the gas industry typically needs large investments to explore and develop reserves and to build up the transport infrastructure. Governments are no longer able to meet the financial needs of the gas sector, as they need to focus scarce resources on other priority public needs. Hence, one of the main motivations for gas-sector reform is to attract private investment and financing. This is also the case in South America.

## Gas-sector structure and reforms in South America

In South America, energy has traditionally been considered a strategic sector, with strong government participation. Until the late 1980s, the region's gas industry was dominated by large state-owned monopolies, generally vertically-integrated with control over the whole oil and gas chain. The debt crisis of the 1980s dramatically reduced the capacity of South American SOE to borrow on international markets. In addition, in many countries, tariff setting and cross-subsidies led to large economic inefficiencies and crippled the ability of state-owned companies to invest their own capital. In the upstream sector, the most visible result was a significant reduction in exploration activity, with a consequent fall in the reserve-to-production (R/P) ratio. In the downstream sector, years of under-investment in transportation and distribution infrastructure, as well as in maintenance, led to a severe deterioration in the quality of service.

After a "lost decade", as the 1980s have been called, the 1990s brought a sweeping change in the region's political and macroeconomic environment, characterised by widespread democratisation and far-reaching structural and economic reforms. These reforms entailed a redefinition of the roles of the state and the market in all sectors of the economy, including in sectors previously considered strategic areas.

In general terms, the first step in reforming the region's gas industries was the financial rehabilitation and restructuring of the SOE operating in this sector. This included adjusting prices to reflect costs and, in some cases, vertical and horizontal break-up of the SOE to facilitate divestment of assets and the entry of new players. New legal frameworks, and in some cases constitutional changes, had to be adopted to allow private participation in sectors that were previously reserved for the state or to permit privatisation of public assets. New laws also established regulations and regulatory bodies to supervise the gas transportation and distribution sectors.

While gas-sector restructuring and privatisation also responded to sector-specific objectives, including the need to improve efficiency and expand the range and quality of services, the main driving force behind gas reforms in South America was the collapse of the previous model of public financing for infrastructure sectors and the need to attract private capital. In some countries, the privatisation of public assets was not so much the result of a coherent and well-developed reform strategy as an expedient way to address macroeconomic problems. In particular, privatisation of state-owned oil and gas companies were expected to:

■ reduce public-sector debt in the short term, as a result of one-off proceeds obtained from the sale of state-owned companies, and in the long term, by leaving the financial burden of the sector expansion to private investors, thus freeing up resources for other uses;

■ improve the country's balance of payments by attracting a steady stream of foreign direct investment;

■ increase the net balance of foreign currency, by raising exports or reducing imports of high-value energy products.[2]

Despite similar driving forces, there are considerable differences among South American countries in the scope and pace of their gas-market reforms, arising in large part from each country's situation in terms of resource endowment (self-sufficient producer or importer); size and characteristics of the domestic market; and historical factors like the existence of large state-owned companies and their level of development and technological experience in the petroleum industry.

Argentina, which has the largest and most mature gas market in South America, led the region's gas reform trend in the early 1990s. The 1989 State Reform Bill provided for the deregulation and opening of the upstream oil and gas sector to private and foreign participation. In 1992, the Privatisation Law allowed the privatisation of the state-owned upstream company *Yacimientos Petrolíferos Fiscales* (YPF) and the Natural Gas Act mandated the vertical and horizontal unbundling and privatisation of the gas transmission-and-distribution monopoly *Gas del Estado* (GdE). The Argentine gas market is currently the most liberalised in the region, characterised by private companies in all segments of the industry – exploration and production, transmission and distribution. Peru and Bolivia followed suit in the mid-1990s, though with different modalities. Peru began the unbundling and privatisation of the assets of *Petróleos de Peru* (Petroperu) in 1993, while Bolivia broke-up state-owned *Yacimientos Petrolíferos Fiscales Bolivianos* (YPFB) and in 1996-97 "capitalised"[3] the resulting upstream and

2. This was the intended case in Argentina, for example. However, it is clear now that the large privatisation programme carried out in the early 1990s did not achieve any of the above macroeconomic objectives. The sale of public companies only provided short-term relief, allowing the government to further delay much needed, but painful, economic reforms. Nevertheless, the restructuring and privatisation of the gas sector did result in an increase in proven reserves, in the expansion of the gas infrastructure and in better service.

3. The word "capitalisation" describes the particular privatisation process chosen by Bolivia. In this process, 50% of state company shares were sold to private investors, who were also granted management control. The remaining 50% of the shares were transferred to a trust fund, with dividends going to a Bolivian pension fund.

transportation units, creating mixed (public-private) companies, controlled by private shareholders.

In Venezuela, Colombia, Brazil, Chile, Ecuador and Trinidad & Tobago, oil and gas companies remain in public hands, but reforms have been carried out to allow or increase private participation in some or all links of the gas chain. In the upstream sector, this has generally been achieved by reallocating to new entrants the exploration and development rights previously held by the state-owned companies (e.g. Brazil); licensing areas not already leased; and by promoting additional joint-venture partnerships (e.g. Venezuela). In the downstream sector, most of these countries have moved towards liberalisation of gas transmission and distribution. Venezuela is the only country where these activities are still 100% in the hands of a state-controlled company, but even in Venezuela there are now plans to award concessions to private companies for the construction and/or the operation of pipelines and distribution networks. *Table 4.1* gives an overview of the current ownership structure and degrees of openness to private participation in the region.

**Table 4.1**  Ownership structure and degree of openness to private participation in South America's natural gas industries

| | Predominantly state systems | | Mixed systems | Predominantly private systems |
| | with limited openness | with total openness | | |
|---|---|---|---|---|
| **E&P** | Venezuela | Brazil, Chile, Colombia | Bolivia | Argentina, Peru, T&T |
| **T** | Venezuela | Brazil | Bolivia, Colombia | Argentina, Chile, Peru, Uruguay |
| **D** | Venezuela | | Brazil | Argentina, Bolivia, Colombia, Chile, Peru, Uruguay |

*E&P = exploration and production; T = transmission; D = distribution.*

The opening of the upstream sector to a larger number of companies has in many cases helped to firm up gas reserve estimates, boost production and lower the development and production costs of domestic gas. Bolivia and Trinidad & Tobago are the best examples of this trend. In Bolivia, the large infusion of private capital in the upstream sector brought a surge of investment in exploration and production, which produced a seven-fold increase in proven gas reserves in just four years.

Gas-sector reforms in South America have generally been part of wider energy-sector reforms, with divestiture of other state-owned energy companies and substantial power-sector reforms taking place before or simultaneously. Indeed, Chile and Argentina were at the forefront of electricity reforms during the 1980s and early 1990s, providing an impulse for changes in neighbouring countries.

# THE ROLE OF THE STATE IN GAS MARKETS

**Policy formulation is a fundamental role of the state**

Governments around the world have all been involved in gas-market development, either through ownership of state-owned companies, regulation of prices, negotiation of gas import projects, bilateral agreements for cross-border pipelines, promotion of gas use through fiscal incentives and other measures.

As gas markets move towards increased liberalisation, the role of government shifts but does not disappear. While commercial decisions on gas supply allocation and prices and detailed investment choices are best left to the market, policy formulation remains a fundamental role of the government. This includes the setting of the country's main energy policy objectives, the formulation of a strategy for achieving them and the establishment of appropriate institutions to implement the strategy. In addition, the government needs to monitor market developments closely in order to be able to address in a timely and effective manner any market failures and externalities.

Some of the key objectives of energy policy are the provision of clean and affordable energy to the final consumers; ensuring long-term energy security; enhancing the economic efficiency and competitiveness of the energy system; and reducing the health, safety and environmental impacts from energy activities. More specific objectives may include promoting energy diversification and energy conservation; supporting the development of domestic energy sources; and promoting more environmentally-acceptable energy sources. In developing countries, ensuring access to commercial energy for all is also a very important objective. In South America, as in other countries with rapidly-expanding economies and energy demand, attracting investment and ensuring financing for energy projects is often another fundamental goal of energy policy.

The government's gas-sector policy objectives, strategy and instruments would normally be embodied in a coherent framework of laws and regulations and lead to the establishment of appropriate institutions to implement the policy.

**Clear separation of government's functions is best**

There is no catch-all, prescriptive approach to the functional organisation of governmental activities in the natural gas sector, but experience around the world has demonstrated the merits of clearly separating responsibilities for policy-making, regulation and – if applicable – dealing with state companies. Assigning each of these roles to different government bodies helps to:

■ prevent collusion between government agencies and state companies, if any;

■ reduce the potential for conflicts of interest within government departments that carry out more than one of the above roles;

■ improve the transparency and reduce the arbitrariness of decision-making, thereby reducing uncertainty, increasing predictability, lowering the cost of capital and increasing the attractiveness of investment;

■ concentrate expertise and competence in each area, leading to more efficient governance.

In most IEA countries, policy-making and regulatory functions are assigned to different government agencies. This is also the case in most South American countries, where ministries of energy are (or should be) in charge of energy policy, while specialist regulatory agencies were created as a result of the reform and liberalisation of midstream and downstream activities to implement the new legal and regulatory framework. Chile is an exception, as policy-making and regulatory functions are carried out by the same body (the *Comisión Nacional de Energía*).

## Dealing with state-owned companies

In the countries where gas companies have remained in public hands, their management creates special challenges. Before reform, state-owned companies were often involved in policy-making and regulation. This is no longer possible in a liberalised market, where the government must set a level playing field for all players. Also, policy-making functions should be exercised by institutions which have a democratic mandate to do so. One of the key aspects of reforms is therefore the elimination of policy-making functions from the state-owned energy companies.[4]

As long as the state retains total or majority control of these companies, it needs to set their strategic, operational and financial objectives. However, the government does not necessarily need to concern itself with the day-to-day management of such companies. Experience around the world shows that it is more efficient to devolve full responsibility for day-to-day management to the companies themselves, governed by a board of directors and a government-appointed chief executive, thus establishing a more arms-length relationship between the government and the companies.[5] There is a move in this direction also in South America, where a clear example of the soundness of this approach is offered by the successful financial results of Brazil's Petrobras, which underwent major transformations and is adapting remarkably well to the new competitive environment.

## THE LEGAL FRAMEWORK

Almost every country with a natural gas industry, whether based on indigenous resources or imports, has adopted a gas law or laws in the early stages of its gas-market development, often amending legislation as the industry matures or in response to changing market circumstances and policy objectives.[6] In most cases, these laws cover the mid- and downstream aspects of the industry. Upstream aspects (where relevant) are usually covered by separate legislation covering both oil and gas.

Through the legal framework, the government defines the rules of the game for all parties involved in the gas chain, providing a clear legal expression of the government's

---

4.  However, state-owned energy companies usually retain significant political power, if only because they are usually among the largest employers in the country.
5.  In this case the government may set specific objectives and performance targets for the state-owned company, eventually covering a period of several years.
6.  For example, laws in North America and all countries of the European Union have been modified in recent years to introduce gas-to-gas competition based on third party access to the network.

policy and strategy for gas-industry development. A clear, unambiguous and stable legal framework is a pre-requisite for attracting investment in the gas sector, as it helps to create a more stable investment and operating environment, reducing uncertainty and investment risk, and consequently lowering the cost of capital. Furthermore, codifying the roles and responsibilities of different players in the industry reduces conflicts of interest.

The legal framework for gas needs to be in line with the country's general energy policy objectives. It also needs to be coherent with other pieces of legislation impacting on the gas sector, e.g. electricity-sector laws, foreign investment laws, environmental laws, capital market laws, fiscal laws, labour laws, etc. This is important since a coherent set of legislation will promote investment, while disjointed and ambiguous legislation will hinder it.

Laws are inherently rather inflexible. Hence, *legislation* will normally establish the overall objectives and framework covering the main rights, obligations and regulatory principles, while authorising the administration to issue more specific *regulations* (or secondary legislation) and, generally, providing for the creation of regulatory institutions.

## Upstream legislation

Upstream laws, usually covering the management of both oil and gas resources, generally include provisions on:

- the ownership of resources;[7]

- the procedures and criteria for awarding rights and licences;[8]

- the licence conditions, including duration and conditions for relinquishment and procedures for modifying rights and licences over time;

- the royalty and fiscal regime;

- the rights and duties of all industry participants (private and state companies) and government agencies;

- access to land and right-of-way issues;

- health and safety and environmental standards; and

- dispute settlement and arbitration.

## Mid- and downstream legislation

The main aspects generally covered by mid- and downstream natural gas laws include:

- broad objectives for the development of the gas industry, including provisions for structural reorganisation, vertical and horizontal break-up, open access to networks, etc.

---

7. In most countries, the state defines itself as the owner of all subsoil resources. One exception is the US, where the owner of the surface land also owns all resources found directly beneath the surface.
8. The exploration phase and the development and production phase are usually treated separately.

■ rights, responsibilities and obligations of the various public and private entities involved in gas transmission, distribution and marketing;

■ general principles and administrative procedures for issuing licences, authorisations and concessions; procedures may also cover approval of long-term contracts for the purchase and sale of gas;[9]

■ allocation of responsibilities to different public bodies for licensing and regulation of the various activities of the gas companies at each stage of the supply chain;

■ tariff principles to be applied by the responsible authorities;

■ issues related to land use and rights of way;

■ health and safety and environmental rules;

■ technical issues, such as standardisation of operating practices;

■ procedures for dealing with disputes.

## Gas legislation in South America

Most South American countries are oil and gas producers. In the 1990s, most of them revised their upstream legal frameworks to reflect policy changes and structural reforms. The new oil and gas laws generally removed legal barriers to the entry of new players and expanded incentives to attract foreign investment, with the objective of increasing exploration and proven reserves, raising domestic production and increasing exports (or reducing imports).

In Brazil, where private investment in the upstream sector was not allowed, the 1995 constitutional amendments and the 1997 hydrocarbons law removed Petrobras's exclusive rights over exploration and production. Petrobras was allowed to keep all of its production acreage as well as acreage where substantial development had already been carried out. However, it had to relinquish a large part of its exploration acreage for reallocation through competitive bidding to all players (including Petrobras itself).

In Bolivia, Colombia, Chile, Ecuador and Peru, the new laws aimed mainly to expand existing incentives for private investment. Incentives generally included measures to allow greater participation in petroleum production (e.g. in Bolivia, Colombia, Chile and Ecuador), permission to freely sell the fuel produced (e.g. in Peru), and reduction in the levels of income tax and taxes on profit repatriation in all the above countries. Additional incentives included more favourable contractual terms and operating conditions, such as the extension of exploration periods; the reduction of the number of exploratory wells to be drilled; freedom to transfer or cede contracts to third parties; and the acceptance of international arbitration to resolve conflicts.

---

9. "Approval" means that the government authority is not involved in the negotiation of contractual terms and conditions, but it has the right to object before a contract takes effect if it judges that any aspect of the contract may work against the public interest.

In Venezuela, a controversial new hydrocarbons law was adopted in November 2001 by presidential decree.[10] Highly criticised both domestically and abroad, the new legal framework is viewed by many observers as actually reducing the incentives and scope for private investment, for example by increasing royalties and requiring that the state-owned company PDVSA hold a majority stake in all oil exploration and production joint ventures.[11]

Argentina is perhaps the only country that lacks a new, comprehensive upstream hydrocarbons law, adapted to the new institutional and legal structure of the sector. A hydrocarbons bill has been awaiting discussion in Congress since 1995, delayed by successive cabinet crises, other more urgent bills and, more recently, by the economic and political crisis.

Most South American countries have enacted or are in the process of enacting mid- and downstream gas laws. Some of these are listed in *Table 4.2*. Those of Argentina, Brazil, Bolivia and Venezuela are discussed in the respective country chapters of *Part 2*.

**Table 4.2** Mid- and downstream gas laws in selected South American countries

| Country | Law | Enacted |
|---------|-----|---------|
| Chile | Law Nº 18.856 – Ley de Servicios de Gas | Dec. 1989 |
| Argentina | Law Nº 24.076 – Ley del Gas Natural | May 1992 |
| Peru | Law Nº 26221 – Ley Orgánica de Hidrocarburos | August 1993 |
| Bolivia | Law Nº 1689 – Ley de Hidrocarburos | April 1996 |
| Brazil | Law Nº 9.478 – Lei do Petróleo | August 1997 |
| Colombia | Law Nº 142 – Ley de Servicios Públicos Domiciliarios<br>Law Nº 401 – Ley de Creación de Ecogas | 1994<br>1997 |
| Venezuela | Decree Nº 310 – Ley Orgánica de Hidrocarburos Gaseosos | Sept. 1999 |

## PRICING AND TAXATION

Adequate pricing of natural gas is key to the successful development of gas markets. Gas has no exclusive markets; hence the ability of gas to penetrate the market depends on its ability to compete with other fuels. The price of gas to final consumers, relative to that of competing fuels, will therefore affect the volume of gas sold and the expected revenues from sales, ultimately determining the viability of any natural gas project. The challenge is to find the right balance between what price the producer wishes to charge and what price the consumer is willing to pay. The pricing of the intermediate stages of the gas chain also needs careful consideration, and each link needs to be priced according to the risks involved.

---

10. Under an "enabling law" which authorised the president to enact a number of particularly important or urgent pieces of legislation, bypassing Congress.
11. This does not apply to exploration and production of non-associated gas, where private companies, domestic and foreign, are allowed up to a 100% stake, but most gas reserves in Venezuela are associated with oil.

Each market has its own unique set of alternative fuels, whose supply, demand and transportation characteristics together set the competitive price for gas. Even within a single market, there may be different prices depending on the type of energy service a customer requires. In South America, gas generally competes with hydropower and other fossil fuels in power generation, heavy fuel oil and biomass in industry (or gas oil and LPG when clean fuel is required), and electricity and LPG in the commercial and residential sectors.

The main question about pricing is how the price is set: whether it is set freely by the market players or fixed by some state authority. International experience has shown that government-set prices distort the economics of the gas supply chain, and fail to give the market the signals and incentives needed to develop the gas market. Market competition is always more effective than any form of government price-setting. When a decision has been made to develop a gas industry relying on private initiative and market forces, pricing decisions all along the gas chain should be left to market players, and the state's role should be restricted to preventing anti-competitive behaviour that could damage consumer interests and to compensating for environmental and other externalities not accounted for in market prices.

## Different pricing methodologies

There are essentially two approaches to gas pricing: the cost-based (or cost-plus) approach and the market-value netback approach, which are schematically illustrated in *Box 4.1*.

In the *cost-plus approach*, the price of gas to the end-user is determined by adding to the price of gas at the wellhead all other elements of cost along the chain, including taxes and a fair return on investment for transportation, storage and distribution. The wellhead price is either set by the producing company based on its costs or regulated by the government.

Under this approach, the economic rent is transferred to the final consumer, instead of being retained by the state or the upstream company. This may be a good way to speed up market penetration of gas. However, as gas is priced independently from alternative fuels, there is no guarantee that the resulting cost-recovery price for gas would allow it to be marketed.[12] Hence, this pricing approach may only work in countries with large and inexpensive gas reserves close to consumption areas.

Even in such cases, this system has drawbacks. The main one is that it does not encourage efficiency improvements and it is ineffective in sending accurate market signals to investors. Added to that, costs are not that easy to determine, as most cost elements are long-term fixed investments, and financing and specific costs are highly dependent on utilisation. Furthermore, costs in a non-competitive environment can easily be flawed. This explains why, even gas-rich countries such as the United States and Canada have abandoned this approach.

---

12. In other words, there is a risk that this pricing mechanism might lead to the development of gas fields that are too expensive for the market; thus requiring some sort of subsidisation.

**Box 4.1**  Cost-plus pricing versus netback pricing

| Cost-plus pricing | Netback pricing |
|---|---|
| Wellhead price (set freely or regulated) | Market value of gas based on price of consumer's competing fuel |
| + Pipeline mark-up cost | − Distributor charges |
| + Local distribution mark-up cost | − Pipeline transportation charge |
| = **Sales price to consumer** | = **Netback price at the wellhead** |

In countries which do not have large and inexpensive gas reserves close to demand centres, a sound gas pricing system has to start with the final consumers' willingness to pay, obviously determined by the overall costs of using alternative fuels – including taxes and a valuation of the environmental benefits of gas. These factors will determine the eventual size of the gas industry: if the price the consumer is willing to pay is not sufficient to finance the development of the necessary infrastructure, then it would be uneconomic to opt for gas.

The starting point of the *market-value netback approach* is the determination of the final consumer's willingness to pay, as expressed by the maximum price of gas at which the end-user would be willing to use gas as an alternative to other fuels. This is the market value or market-replacement value of natural gas for a determined consumer. The value of gas at any point of the gas chain before the final consumer is then calculated "backwards" by deducting from the weighted replacement value of all customers in a given area, the costs (for distribution, load and quality management and transportation) necessary to bring the gas from the relevant point in the gas chain to the final consumers (*Box 4.2*).

In the case of gas use for power generation, which is the most relevant for emerging gas markets, the competitive pricing of gas is complicated by the fact that often natural gas does not compete directly with alternative fuels, because the competing way of generating electricity (e.g. a coal-fired power plant or a hydropower plant) will often require a completely different investment. In this case, competition is not at the burner tip, but at the feeding point to the grid. In other words, the market-replacement value of gas cannot be calculated by just looking at the price of competing fuels, but has to take into consideration the total cost of generating electricity with gas and with alternative energy sources, including the necessary power transmission investment to bring the electricity to the market. However, the evaluation is very different depending on whether it is based on the long-run or on the short-run marginal costs. A gas-fired CCGT or gas turbine will require a substantially lower capital investment than, say, a hydropower plant, but its operation and fuel costs will be higher. Hence, the long-run marginal cost of gas-fired power may be clearly lower than that of its alternatives, but on a short-run marginal cost basis the alternatives may be cheaper.[13]

---

13. This is further complicated by the different incidence that risk has on the various elements of costs for different power stations. Given the shorter lead times and amortisation periods of gas-fired plants, the risk premium on the financing of a gas-fired plant is generally much lower than on hydropower plants. However, in the case of a gas-importing country, the effect of exchange rate risk on fuel price may significantly affect the economics of a gas-fired power plant.

**Box 4.2**    The market-value netback approach

The market value or market-replacement value of natural gas for a determined consumer is defined as the price of gas at which the end-user incurs the same costs when using gas instead of alternative fuels, taking into account differences in heat value, efficiency and costs for the appliances or equipment necessary to use the energy, plus eventually the differences in costs of meeting environmental standards.

The market-replacement value (or market value) of gas thus must take into account the following factors:

- Price and heat content (usually in net calorific value) of the competing fuel(s);

- The "premium" of gas compared with other fuels for the same application, including its higher efficiency, lower operating and maintenance costs, and lower investment requirements.[14] In addition, lower costs to meet environmental standards may be factored in. Eventually, the premium should also internalise environmental externalities, to the extent that they are not reflected in emission or pollution standards.

- Other advantages that are not easily expressed in cost figures, including clean combustion properties, flexible and comfortable application, and security and ready access to supply.

The weighted average market-replacement value of gas for all consumer categories in a given area,

*minus*   the costs incurred in the distribution sector to bring natural gas from the city-gate to all consumers, taking into account storage and other load management costs;

*minus*   the taxes on the use/distribution of gas;

*equals*   the netback value at the city-gate.

The netback value at the city-gate,

*minus*   the costs of transporting and storing the gas from the wellhead or LNG terminal or import border to the city-gate;

*minus*   the taxes on the transportation and imports;

*equals*   the netback value at the wellhead, at the import border or at the LNG regasification terminal.

This approach, by definition, involves price discrimination between different categories of end-users (or even between gas customers of a same category but in different geographical locations) because each of them will have a different gas demand profile determined by the practicality and cost of using alternative fuels.

---

14. Negative premiums, or disadvantages, should also be factored in, e.g. the higher storage costs of gas compared to alternative fuels.

The market-replacement netback value typically serves as a basis for price negotiations between market players at any point in the gas chain. In the case of captive customers, the replacement value may serve as a benchmark for authorities to set regulated tariffs or to assess *ex post* if a distribution company abuses its dominant/monopoly position (see *Box 4.3*).

**Box 4.3**          Price regulation for small consumers

Small consumers, such as households and small commercial users, are generally captive customers, in that they have no choice of supplier and/or a very limited ability to switch fuels (in most cases, it is not economic or practical for small consumers to maintain dual-firing equipment or back-up fuel supplies). Most countries regulate prices to these consumers to prevent an abuse of market power from the local distributor.

From an overall economic point of view, prices/tariffs to captive customers should be limited by the market replacement value of gas, with due consideration of possible premiums from environmental advantages, which are not easy to calculate. On the other hand, there seems to be no rationale to stay below such a price level as long as the gas sector can still be enlarged to bring its environmental advantages to more people. To achieve optimal expansion of the gas market, it should be ensured – e.g. by taxation schemes – that extra revenue for the companies in the gas chain is invested in the development of the gas industry.

In most countries, there are explicit controls on the prices of bundled gas sales (the cost of the gas itself plus the cost of delivery) to small customers. In Britain, where competition was extended to all customers, including households, in 1998, delivered gas prices are no longer regulated, but national transmission and local distribution charges, which are included in delivered gas tariffs, continue to be regulated on a rate-of-return basis. Germany is the only country in Europe where there are no explicit regulatory controls on gas tariffs for small consumers, but national competition legislation prevents companies from exploiting market power and charging excessive prices. In addition, the large number of local distributors owned by the municipal authorities reduces the threat of monopoly abuse.

The netback market value may in some cases be significantly higher than the cost of gas production and supply (or the cost of gas at the border or at the LNG regasification terminal). In general, near-to-market producers have an inherent advantage because transportation costs are much lower. As a result, a considerable economic rent may be earned between the average netback market value and the supply cost.

The replacement-value netback pricing approach does not provide a rent for the final consumer. With the netback principle, any rent is passed up the chain to the gas producer. However, negotiations typically result in the sharing of the rent between the producer/producing country[15], any transit country/company, the transmission and distribution companies, and the consumer. As long as there is a rent, parties can negotiate prices below the netback value to speed up market penetration and increase overall sales volume, leading to better economies of scale in production and transportation.

One basic requirement for netback pricing to work is that the prices of competing fuels must also be market-based and undistorted. The government can, however, improve the competitiveness of gas *vis-à-vis* alternative fuels for environmental or other policy objectives. Enforcing stricter emission standards, charging higher tax rates on more polluting fuels, or eventually banning the use of more polluting fuels in certain applications or in certain areas, will have the effect of increasing natural gas's replacement value *vis-à-vis* other fossil fuels.[16] Other policy objectives may also come into play; for example, the government may want to improve fuel diversification and security of supply by increasing the share of gas.[17]

In Continental Europe, a region that depends largely on imports, the netback market-value approach has traditionally been the basis of gas pricing throughout the gas chain. Such a system developed because of the need for the gas companies to recover the large capital costs involved in building the pipeline infrastructure. By pricing gas through the chain in relation to competing fuels, gas sales and per unit revenues are maximised. The price risk and reward – which are linked to the movement of prices for competing fuels – are effectively transferred to the producer, who thus gets any economic rent available from the difference between the netback value and the costs of exploration and production, and transportation of the gas to the delivery point. The risk pattern incurred by the gas producer and the country owning the gas resources is thereby similarly structured to the risk pattern in oil production, the main investment alternative for most gas-producing companies and governments.

In South America, Argentina adopted a netback pricing approach after de-controlling wellhead prices in the early 1990s. The price of Bolivian gas imports into Brazil is also broadly based on a market-value netback approach, as it based on a formula that links the price to a basket of fuel oil prices. However, the price is set in US$ and exchange rate variations are accounted for in the formula. Thus, the recent large depreciation of the Brazilian currency has resulted in a considerable increase in the price of gas compared to alternative fuels.[18] For domestic gas, in countries other than Argentina, the current gas pricing approach is predominantly a cost-plus one.

15. The part of the rent going to the producing companies depends on the upstream taxation and royalty scheme of the producing country. Such risk and rent sharing is subject to international competition, as producing companies will have alternative investment opportunities in other countries in oil or gas and will compare the expected risks (geological, market and political) and revenue on an international scale. On the other hand, many companies are looking for investment opportunities in upstream exploration and production.
16. For example, banning the use of high-sulphur fuel oil from certain applications would result in competition only between gas and low-sulphur fuel oil or lighter oil products, which are more expensive than high-sulphur fuel oil. Hence, the replacement value of gas would be increased.
17. For instance, if energy security in relation to increasing oil import dependency is a key concern, imposing a higher excise tax on oil products than on gas would increase the market replacement value of gas *vis-à-vis* oil products and reduce oil consumption.
18. Admittedly, fuel oil prices, which are now liberalised, should have increased by a similar factor. But in Brazil imported gas is mainly used (or expected to be used) in the power generation sector, where it competes mainly with hydropower.

**Taxation**

Taxation of natural gas and other fuels is a key element of a coherent energy policy, and can play a critical role in stimulating the supply of and demand for natural gas, especially in the early stages of industry development. Taxes are not only the main source of fiscal income, to be spent for the general functions of a state, e.g. to support education, defence, the environment, health, the judiciary system, administration, etc, but can also serve as powerful incentives or penalties to achieve a desired pattern of consumption, especially between substitutable products. Taxes can also be used as instruments to internalise externalities and to convert government policies into economic signals to the market.

Many countries have used favourable taxation of gas as a primary means of supporting market development (*Box 4.4*). Using taxes to implement government policy by influencing commercial decisions by private actors is legitimate as long as the framework is clear, does not discriminate between the actors in the market and is not changed retroactively. This use of taxes is especially justified if differences in taxes reflect differences in environmental externalities and costs, which are not accounted for in market-based prices.

Examples of fiscal measures that foster the development of a gas industry by creating additional volume of demand and/or by increasing investments in the gas industry are:

■ reducing taxes on gas use and giving tax incentives to gas consumers for installing gas appliances and equipment;

■ setting lower excise taxes on gas than on competing fuels;

■ providing tax incentives for re-investment of profits into the gas chain;

■ providing tax exemptions for gas-related investments;

■ reducing or removing taxes on imports of pipeline materials and equipment,[19] which can raise significantly the cost of building transmission lines and local distribution networks;

■ replacing volume-based or revenue-based taxes with profit-based taxes, to encourage the development of less profitable projects (e.g. marginal fields);

■ removing differences in tax treatment between domestic and foreign companies.

Governments wishing to promote investment in their gas industry should also:

■ periodically review their fiscal terms to ensure that they remain internationally competitive and are consistently applied to all industry participants;

---

19. When these materials and equipment are not produced locally.

■ streamline the tax system and improve the co-ordination between the central government and provincial and local authorities to avoid the cumulation of taxes;

■ improve the transparency of the tax administration.

Close co-operation between the fiscal authorities and the energy policy institutions is essential to design a taxation regime that appropriately reflects the energy policy objectives of the country.

**Box 4.4**    Example of successful promotion of rapid development of a gas industry

The rapid penetration of gas in the Spanish energy balance demonstrates the importance of price competitiveness and taxation policy. Here, lower taxes on gas sales than on competing fuel oil played a major role in making gas competitive against other fuels and in stimulating rapid fuel switching, as well as in giving the sellers more revenue for investment. Gas consumption increased rapidly since the 1970s, reaching 12% of total primary energy supply by 2000. Originally only LNG was consumed, but this is now supplemented by piped gas from Norway and Algeria. The Spanish government promoted the uptake of natural gas in industry through explicit oil-related pricing and lower taxes on gas (no excise tax) than on oil products (e.g. an excise tax of 40% is levied for light fuel oil). Gas has been consistently much cheaper on a heat equivalent basis than gas oil, and broadly in line with heavy fuel oil prices in recent years. The non-price advantages of gas over fuel oil mean that gas is usually the preferred fuel when an oil or coal boiler is replaced, or for new boilers or direct heat applications.

## THE REGULATORY FRAMEWORK

**Why regulate the gas sector?**

There are four main areas where the state may want to set a regulatory framework for the gas sector.

*(i) Technical and Health, Safety and Environment (HSE) regulation*

In industries like the oil and gas industries, which have a potentially large impact on health, safety and the environment, the state has a clear role in establishing technical standards and monitoring their implementation. Technical regulation must ensure that market participants are technically qualified to undertake the business, and that the infrastructure is built according to appropriate standards to protect the health and safety of workers and customers, and the environment.

*(ii) Use of public land and public streets*

Gas production, transportation and distribution are linked to a fixed infrastructure that uses public land and public streets. Hence, the government needs to define rules and eventually charge fees and taxes for the use of public land and public streets.

*(iii) Resource management and rent sharing*

Upstream activities involve the use of non-renewable resources which are typically considered as part of the national assets and which generate an economic rent. Hence, governments normally establish rules and compensations regarding resource management and depletion. Risks and rewards need to be shared according to clear rules, by the government and the companies participating in the exploration and extraction. This is generally achieved by granting the companies extraction rights against the payment of royalties or extra petroleum taxes, over and above the normal corporate income tax.

*(iv) Market imperfections and externalities (economic regulation)*

In view of the strong market imperfections and externalities that characterise gas markets, economic regulation is needed to give the right incentives to market participants and to protect consumers against the abuse of dominant market power. The main gas market imperfections and externalities to be addressed by economic regulation are:

■ *Natural monopolies*: Grid-based energy supply activities, such as gas distribution and transmission, have traditionally been considered natural monopolies. The supply of a given commodity is a natural monopoly when economies of scale are such that the costs of supply for the total market are lower if there is a single supplier. This is generally the case for gas distribution, as a duplication of the network by a newcomer would, in the vast majority of cases, be a loss-making enterprise and would be economically inefficient.[20] High-pressure transmission pipelines, however, are not necessarily a natural monopoly, although they do involve large economies of scale and high specificity.[21] Indeed, in a sufficiently large market with a large number of suppliers, gas transmission can be a competitive activity, as a given pipeline may be economically challenged by building a parallel pipeline or by a pipeline on an alternative route.[22] However, in small or incipient gas markets, the economies of scale of a pipeline are very large compared to the relevant transportation market and the scope for building a second pipeline over the same route or a competing pipeline from a different supply source will be limited. Given these characteristics, prices charged for distribution and, in some cases, transmission services need to be controlled to prevent the monopoly service provider from exploiting the potential for excessive profits (see *Box 4.3* above). Storage may also display natural monopoly characteristics, though geology and market size may allow storage services to be provided efficiently by the market.

■ *Anti-competitive behaviour by gas suppliers*: There may be concerns about the level of market concentration. A dominant company may seek to stifle competition through predatory pricing, especially in the early stages of competition in gas wholesaling and retailing. In the transition from a former state-owned monopolistic industry to a competitive

---

20. As well as physically impossible due to scarcity of space underneath public streets.
21. Specificity means that once laid down, pipelines can only be used for the purpose they were built for and there is little chance to recoup the investment otherwise.
22. Experience in the US, Canada and Germany shows that in a sufficiently large market, the freedom to build and operate pipelines – and thus the existence of parallel transmission pipelines – is not economically inefficient.

model, regulation is also needed to give new market entrants a fair chance against (former) state-owned companies.

- *Supply security issues*: Security of gas supply can be divided into three risk categories: the short-term risk of disruptions to supplies through the failure of markets to balance supply and demand adequately, the long-term risk that inadequate investment will be made to secure future supplies (noting that short-term security ultimately fails if long-term investment fails); and the risk of inadequate diversity of supply sources in the event of a major disruption from a given source.

- *Consumer welfare protection*: Most governments seek to provide protection to household customers, particularly the poor and disadvantaged.

While there is little difference between countries on technical regulation, the approach to economic regulation varies enormously across countries, reflecting differences in policy objectives, political traditions, institutional and structural factors. Regulation varies considerably from a light-handed or hands-off approach to one characterised by detailed and onerous constraints on gas company activities. The light-handed approach, as in Germany and New Zealand, places primary reliance on general competition law and anti-trust institutions to address anti-competitive behaviour in an *ex post* manner. More detailed regulation, as in the United States, Britain and Argentina, reinforces the foundations provided by competition law with explicit mechanisms to control the behaviour of natural monopolies, including pricing and handling of network access as well as financial and operational performance.

The type of regulatory framework needed also differs significantly depending on the maturity of the gas market. The regulatory framework currently in use in North America, Western Europe and in Argentina is adapted to countries with a well-developed gas industry and markets. It may be premature to adopt a similar one in other South American countries with incipient gas markets.

## Key issues to be addressed by economic regulation

Key elements of the regulatory regime, which may differ significantly across countries, include:

- access to networks;

- controls on price or rate of return for non-competitive segments;

- structure of the sector / unbundling requirements;

- regulatory responsibility.

## Third party access regime

Third Party Access (TPA) is the right of any third party (either a producer, a consumer, a shipper or a trader) to access/make use of the transportation and related services of a pipeline company for a charge (tariff). It entails an obligation for the latter to offer such services, although the extent of such rights and obligations may be limited by relevant legislation. TPA therefore differs significantly from voluntary access, which can take place freely without government intervention.[23] If the economies of scale are large and relevant compared to the size of the transportation market, TPA is the usual method for introducing competition in the transportation segment, by making the transportation accessible for newcomers under the same conditions as the incumbent. TPA may apply to either or both gas distribution and transmission. It may also apply to LNG terminals and storage infrastructure.

There are essentially two types of access regime: regulated and negotiated.[24] In a regulated TPA regime, public authorities (usually the regulator) impose explicit rules on how pipeline companies handle requests for access to the network, and set operational and financial conditions for the use of the system. Under negotiated TPA, on the other hand, market players are free to negotiate the terms and conditions of access. The government may nonetheless play a role in ensuring that negotiated access is effective. For example, it may require the industry to prepare a formal code, setting out essential commercial terms and conditions, including grounds for refusing access. It also normally provides an arbitration mechanism in the event of dispute. TPA, negotiated or regulated may be backed up with requirements for unbundling and information disclosure. Countries may choose one regime for transmission and another for distribution.

The access regimes in North America and Britain are best characterised as regulated TPA. Countries of Continental Europe are just starting to introduce TPA following the adoption of the EU Gas Directive in 1998. The Directive mandates TPA but leaves the choice to each country to decide whether to implement regulated or negotiated TPA. Most countries have opted for regulated TPA.

While TPA works well for existing networks where the investment is in large part amortised, there is an important caveat related to new pipelines. The investor in a pipeline will obviously take some risk, which should either be reflected in his having the exclusive right to use his own investment (at least for some time) or by a risk addition to the tariffs for those using an already existing pipeline to compensate for the risk taken by the original investor. Otherwise, investors might be deterred from investing in new pipelines because of the asymmetry of risk and reward.

An alternative way to introduce some competition in the transportation segment while taking into account the need to protect the interest of the original investor is through an open season scheme. Under this scheme, any pipeline project has to be published and interested companies can join the pipeline consortium before construction starts and will get a reserved portion of the pipeline capacity. The pipeline is usually built with some extra capacity which will be subject to TPA.

In emerging gas markets, when the focus is to provide strong incentives for investments in new infrastructure, it would seem appropriate for governments to offer a degree of

23. In 1992, the European Commission defined TPA as "a regime providing for an obligation, to the extent that there is capacity available, on companies operating transmission and distribution networks to offer terms for the use of their grid, in particular to individual consumers or to distribution companies, in return for payment".
24. A detailed discussion of the characteristics and pros and cons of negotiated and regulated TPA can be found in IEA (2000).

**Box 4.5**        General conditions and issues to be considered for third party access

Regardless of whether TPA is regulated or negotiated, there are certain conditions for the application of such a regime:

- There should be a sufficiently well-developed gas market with excess pipeline transmission capacity;

- There should be a sufficiently large number of gas producers and consumers who seek to have access to the spare capacity rather than building their own pipelines; and

- Physical links should exist or be feasible between existing pipelines.

The TPA regime should ensure access to all network users (or to a defined class of customers, known as "eligible" customers), be they individuals or companies, on equal conditions. The main issues be addressed by a TPA regime are:

- Eligibility for participation, i.e. which market players should be able to benefit from TPA; for instance, whether players should be of a certain minimum size (if gas consumers) or should meet technical and financial standards (if shippers);

- Definition of the facilities to which access is to be granted: e.g. transmission and distribution pipelines both onshore and offshore; LNG facilities; gathering, storage, treatment and blending facilities;

- Definition of the services that may be involved apart from the transmission: e.g. metering, pressure balancing, quality management, load balancing, storage, back-up and stand-by services;

- Determination of the extent to which the pipeline company is obliged to provide these services separately; i.e. to what degree these services should be unbundled;

- Definition of available capacity and the procedures to be followed when capacity is not sufficient (queuing procedures, requirements to build capacity);

- How to determine the tariffs for transportation and related services;

- How much information the pipeline company will be required to disclose regarding availability of, and calculation of charges for, services;

- Technical rules and standards to ensure operability.

Further issues to be addressed by policy-makers are:

- The regulatory framework (regulatory bodies, instances of appeal, etc.);

- Legislation dealing with transitional problems caused by the introduction of TPA;

- Settlement mechanisms to ensure expeditious resolution of disputes;

- Mechanisms to avoid abuse of dominant positions.

protection to investors in high-pressure transmission pipeline, local distribution networks, storage facilities and LNG terminals. Transmission, LNG or distribution companies should not be legally obliged to offer transportation or regasification services to third parties (at least for a given period), although the companies would be free to negotiate such services if they wish. The EU Gas Directive, for example, establishes derogations to TPA requirements for those countries, or areas of countries, which qualify as an emergent market. Derogation may be granted for up to ten years after the first gas supply in the country or in the area.[25] The introduction of a TPA regime to encourage gas-to-gas competition should nonetheless remain a longer-term objective, and the government should make its intention and timetable for introducing TPA clear to investors.

The question is different for access to existing infrastructure, especially that of state-owned or formerly state-owned companies. Here the infrastructure was usually not built at an investor's own risk, but was supported by state financing. The arguments developed above for new pipeline infrastructure do not hold in this case, and given the dominant position of state-owned companies or formerly state-owned companies it is usually necessary to ensure TPA to existing infrastructure to allow entry of new market players.

In any case, wherever possible, enabling competition in building new pipelines should be a priority. If a new pipeline can be built economically, there is *a priori* no reason why the incumbent gas pipeline operator should build it, rather than another operator.[26] To enable competition, the government should not give preference to state companies in assessing applications to develop new transmission pipeline projects. It should also remove the *de facto* monopoly rights over gas transmission that state-owned companies currently enjoy. This would enhance incentives for companies to lower costs and improve the quality of service.

In South America, Argentina has chosen regulated TPA for both transmission and distribution, while Brazil has opted for negotiated TPA for transmission only. An example of a pipeline built under an open season scheme is the GasAndes pipeline linking Argentina and Chile. The consortium includes gas producers, a gas distribution company and a power generator. The expansion of the Bolivia-Brazil pipeline is also planned to be done under an open season scheme, with limitations for the owner of the existing capacity.

**Controls on price and/or rate of return for non-competitive segments**

Regulation does not only have to address the rules for TPA, but also the issue of remuneration for the capacity used by a third party.

Most countries maintain some form of explicit price controls, either on bundled gas supply or on unbundled transmission, distribution and storage services. These controls are justified by the potential for a natural monopoly company to earn substantial

---

25. It also gives derogations to TPA to countries which are dependent on one main external supplier (with a market share of more than 75%) and to those which are not interconnected to another Member State. Gas utilities can also apply for derogations to TPA if they would encounter serious economic or financial difficulties in relation to existing take-or-pay contracts.
26. The incumbent, however, has the advantage of not having to negotiate and pay for new right of ways, as it can lay the second pipeline right next to the first.

economic rent and the lack of incentives for the monopolistic enterprise to operate efficiently. Such controls may be based explicitly on a calculation by the regulator or other government authority of an appropriate rate of return for the pipeline company. Where competition has been established in gas supply, price controls only apply to the transmission and distribution business.

While the costs of new pipelines are relatively clear-cut and the book value is usually close to the replacement value, the issue is more complicated for older pipelines, especially if they have been devalued by high inflation. In that case, the book value is much lower than the replacement value, and tariff setting will either pass on an undue benefit to a newcomer – if based on book value – or it will create a large profit for the pipeline owner – if based on a replacement value.

Controls on bundled gas-supply prices may be retained provisionally for the dominant supplier until competition has become well established, as in the British residential sector. Some countries, such as Germany, do not explicitly or directly control or regulate prices on the grounds that inter-fuel competition effectively limits the prices that gas companies can charge throughout the gas chain.[27] In New Zealand, there are no explicit ceilings on price or rate of return, but the government may impose price controls on gas companies found to be abusing dominant market position.

In Argentina, the regulator sets maximum tariffs ("price caps") for transportation and distribution on the basis of the cost of providing the service plus a reasonable rate of return on investment. Transmission and distribution companies are free to offer lower tariffs to their clients. However, to ensure transparency and non-discriminatory treatment, they are required to publish their tariffs for different services and customer categories. The price caps are adjusted for inflation every six months and reviewed by the regulator every five years, taking into account efficiency improvements and investments.

**Structure of the sector / unbundling requirements**

The other main issue governments have to address is the structure of the sector, i.e. to what extent vertical and horizontal integration and cross-ownership along the gas chain or with other sectors, e.g. the electricity sector, should be allowed. No single answer or model fits all situations. The appropriate choice for a country will depend on its specific circumstances, in particular on the degree of maturity and the overall size of the market, and on whether the country is a producer or importer of gas.

As noted earlier, a key challenge with TPA is to secure non-discriminatory treatment of access seekers and in particular to ensure that an integrated transport company does not discriminate in favour of its own gas supply business. When the owner of the gas pipeline also has interests in production and distribution, it has the ability and the economic incentive to restrict competition in both activities by denying or restricting access to the pipeline to other producers, or by raising the price or lowering the quality of the services it provides to other companies in comparison to those it provides

---

27. Nonetheless, the Federal Cartel Office in Germany recently negotiated reductions in distribution company tariffs which were deemed to be significantly above the national average. The Cartel Office resorted to court action and the threat of legal proceedings under German competition law.

to its own gas supply business. The diversification of upstream activities, as well as true consumer choice, can realistically work only when all producers can have non-discriminatory access to pipeline transmission to sell their gas.

Unbundling, i.e. the separation of a vertically-integrated company's transportation business from gas supply and trading activities, may avoid self-dealing or other forms of discriminatory behaviour. Unbundling also aims at ensuring that costs are correctly allocated to a gas company's different activities, thereby limiting the scope for cross-subsidisation and facilitating the task of regulating transmission and distribution tariffs.

Unbundling can take different forms. From the weakest to the strongest requirement, these are:

■ *Accounting separation*, i.e. the requirement for the integrated company to prepare separate accounts for its upstream, midstream and downstream activities. The company should charge itself the same prices for transport services as it does others and should identify prices for the commodity, for transport and for ancillary services.

■ *Management or functional separation:* i.e. the requirement for companies involved in various segment of the gas industry to set up separate, distinct divisions or subsidiaries.

■ *Operational separation:* operation of, and decisions about, investment in the transport system are the responsibility of an entity that is fully independent of the gas merchant; ownership of the transmission grid remains with the gas merchant.

■ *Divestiture or ownership separation:* gas sales and transport are separated into distinct legal entities with different management, control, and operations; and there is no significant cross-ownership.

Ownership separation solves most concerns because it eliminates both the incentive and the ability to discriminate. It also ensures that each company focuses on its core activity, and allows costs to be better allocated to specific functions. Restrictions to cross-ownership, which have been used in the gas and electricity sectors in several countries, such as Argentina and the United Kingdom, has been shown to be effective in ensuring a level playing field in marketing gas and electricity to local distribution companies.

The "weaker" forms of unbundling reduce, to different extents, the ability to discriminate and may be easier to adopt in some countries, but they do not eliminate the incentive to engage in discriminatory behaviour as effectively as ownership separation. Additional measures to ensure non-discrimination, such as the establishment of "Chinese walls", may be required.

One problem with ownership separation is that it may lead to inefficient investment decisions. Vertical integration offers economies of scope, helps investment synchronisation along the gas chain, minimises transaction costs and reduces the risks associated with the weakness of contractual enforceability. For these reasons, during the early stage of gas market development, when investment in the development of gas reserves and infrastructure is closely tied to specific markets being established, a vertically-integrated structure is perhaps the most appropriate. Setting limitations on vertical integration and cross-ownership at this stage could reduce investment and slow the development of the gas market.

Prior to reforms, all countries in South America had high degrees of vertical integration: either one company was responsible for all activities along the gas chain or, in some cases, two companies existed, with transmission being bundled with either the upstream or the distribution segment.

Today the situation is far more varied. In Argentina, the law expressly prohibits vertical integration and severely limits cross-ownership along the gas chain. No single player may hold a controlling stake in a company operating in another segment of the gas chain. Transport companies may neither purchase nor sell gas. The same is true in Bolivia, except for projects and operations in remote areas and activities involving the development of new domestic markets which would not be financially viable in case of vertical disintegration. Colombian law, too, calls for a total separation of gas production, transmission and distribution activities, but implementation is gradual.

At the opposite end of the spectrum is Chile, which does not expressly prohibit vertical integration. Indeed, the state-company ENAP carries out production, transportation and distribution in the southern tip of the country where gas reserves are located. Even in the case of gas imported from Argentina, transmission and distribution are carried out by private consortia, with substantial cross-ownership.

Brazil and Venezuela opted for functional separation, requiring vertically-integrated oil and gas companies to set up separate, distinct divisions or subsidiaries to carry out transmission and distribution activities. There are no limits to cross-ownership and indeed both Petrobras and PDVSA have fully-owned subsidiaries which own and operate their transmission assets

**Regulatory responsibility**

The regulatory framework that has been described above needs appropriate regulatory institutions to manage it. In most countries around the world, a specialist agency or authority is responsible for implementing the regulatory framework. The authorities' objectives and *modus operandi* are laid down in legislation. Responsibility may be vested in an individual, as in Britain, or in a commission made up of a small number of individuals: five in the United States and Mexico, and three in Italy. Among IEA countries, only New Zealand and Germany have no sector-specific authority and rely *ex-post* on anti-trust authorities to investigate allegations of anti-competitive behaviour under general competition law. Despite significant differences in the approach to regulation, a number of general principles are commonly regarded as desirable in regulatory practice (*Box 4.6*).

Most South American countries now have a regulatory agency dealing with natural gas. In Argentina, it is the *Ente Nacional de Gas* (Enargas), in Bolivia the *Superintendencia de Hidrocarburos*, in Colombia the *Comisión de Regulación de Energía y Gas Combustible* (CREG) and in Venezuela the *Ente Nacional del Gas* (Enagas). In Brazil, upstream oil and gas regulation is carried out at federal level by the *Agência Nacional do Petróleo*

(ANP). Downstream regulation, however, is the responsibility of the states, most of which have now created state regulatory agencies, often covering several service industries as well as gas distribution.[28] In Chile, regulatory responsibilities for gas are carried out by the *Comisión Nacional de Energía* (CNE), which also has a policy-making role.

## Box 4.6      General principles of regulation

Regardless of the chosen approach to regulation and detailed regulatory arrangements, a number of general principles should characterise the regulatory system. These include the rule of law, transparency, neutrality, predictability and consistency, accountability and independence:

- The *rule of law* is the foundation of a regulatory system as it ensures the legitimacy of regulation. Decisions of the regulator should be subject to appeal within the judiciary system.

- *Transparency* is essential for regulatory quality. Transparency involves the capacity of regulated entities to express views on, identify, and understand their obligations under the rule of law. It is an essential part of all phases of the regulatory process – from the initial formulation of regulatory proposals, to the development of draft regulations, through to implementation, enforcement, and review and reform, as well as the overall management of the regulatory system. Public consultation and accessibility are two key instruments for improving transparency.

- *Neutrality* means that the regulations should be neutral to all market players without favouring one or another group (non-discrimination).

- *Predictability and consistency:* Rulings and judgements issued by the regulatory authority should be consistent and should have a reasonable degree of predictability based on previous rulings in similar cases.

- *Independence* of the regulator from the regulated companies is a prerequisite for any sound regulatory system. Independence from government and political actors in the implementation of legislated policy may be desirable to ensure long-term stability of regulatory policies. This independence is critical in countries where there is public ownership of gas utilities. Independence also implies that the regulator needs to be provided with the adequate resources, skills and information.

- *Accountability:* Independence of the regulator must not be confused with lack of accountability. Regulatory agencies, like any other public body, must be held accountable for their actions and must be subject to adequate efficiency controls, especially in those areas not directly related to gas (e.g. general management).

---

28. While the ANP is well resourced and well staffed, most of the state regulatory agencies have just been created and lack specialised gas expertise.

In a growing number of IEA countries, responsibility for regulating the gas and electricity industries is combined in a single body to keep pace with the trend of convergence between gas and electricity industries, as gas-fired power generation plays a growing role in a new gas market and there are clear synergies in the joint supply of gas and electricity to the end consumer. This trend has not yet reached South America.

South American governments will need to decide whether the pace of development of competition in both gas and electricity markets (and the dependence of the gas sector's development on the power sector) warrants the creation of a single agency covering both sectors, or whether to keep two separate agencies (putting in place appropriate co-ordination mechanisms). Independence from electricity regulation may be important, at least in the beginning, to ensure that the specificities of the natural gas sector are properly taken into account. Alternatively, if a joint electricity/gas body is set up, it is important to ensure that it has the appropriate expertise to deal with gas.

## Specific issues related to cross-border gas projects

Physically, building a cross-border gas pipeline is like building two gas pipelines in different countries at the same time. Commercially and operationally, however, it is more complicated. A cross-border pipeline involves international energy trade, and if it passes through a third country, transit issues also have to be resolved.

A cross-border pipeline involves more players than a domestic project, including: the producing country; the company(ies) contracted for the exploration and production of gas; the transit country, if applicable; the transport companies in the producing, transit and consuming countries; the consuming country; the purchasing company in the consuming country; the distribution companies in the consuming country, and the final customers. When more than one sovereign state is involved, there is a need for all the states to share the risks and rents. Alignment of interests of all parties is the key. This needs to be done in a way that ensures a well-balanced allocation of risks and returns for all players.

Cross-border pipeline projects must also be submitted to regulatory review simultaneously in all the countries involved. The regulatory risks will be reduced if the technical and HSE regulatory framework and procedures are predictable and reasonably efficient in all countries involved. Similarly, a clear economic regulatory framework that is stable over time will enable project sponsors to make their investment decisions based on a reasonable assessment of the commercial merits of a project.

Cross-border projects do not necessarily need full harmonisation of regulation or similar degrees of market liberalisation on each side of the border, but some degree of co-ordination between the countries involved is needed, for example to ensure that investors are not taxed twice (e.g. via double taxation agreements).

In the longer term, integration of gas markets would be facilitated by harmonisation of:

■ energy taxation, including the levy of royalties and concession fees;

■ access and tariff regulation;

■ environmental standards and regulations;

■ technical standards, specifications and practices.

This is however likely to take a long time (judging by the European experience). In the meanwhile, governments should focus on creating the right conditions for investment in their own country, establishing clear energy and gas policy, as well as appropriate and stable regulatory frameworks. In addition, governments can facilitate understanding between sellers, buyers and transit countries through regional co-operation frameworks or regional fora. The acceptance of international investment and transit protection rules may also provide additional guarantees to the parties involved.

## CRITICAL POLICY ISSUES FOR SOUTH AMERICA

The circumstances applying to the gas sectors of the countries of South America are as many and varied as the countries themselves, and so are the opportunities and challenges faced by each country. Hence, this section does not attempt to make specific recommendations for each country. Rather, it aims to raise some of the key policy issues that South American countries might need to address in order to foster gas market development.

**Addressing macroeconomic issues**

Macroeconomic and energy issues are tightly intermingled. On the one hand, the provision of affordable, clean and secure energy is a necessary precondition for sustained economic growth. On the other hand, macroeconomic performance has a significant influence on gas-sector development. Economic growth influences the rate of growth in gas demand, which is one of the main factors that determine the economic viability of gas projects. Investments in gas projects are also very sensitive to economic and financial stability, as they involve large capital outlays with long recovery times. Hence, sound economic policies are the necessary prerequisite to promote gas-market development.

**Strengthening energy policy-making capacity**

As countries move towards increased private-sector participation in their gas sectors, the role of the state shifts but does not disappear. While commercial decisions are best left to commercial players, the government has the responsibility to set an adequate policy, legal and regulatory framework to attract private investors, while ensuring that its policy objectives are met. Energy policy formulation is an essential role of governments and parliaments.

In South America, most countries have ministries that deal specifically with energy, reflecting the high importance of energy in energy-rich countries and in countries whose energy demand is still growing rapidly. Often, however, insufficient resources and high turn-over in the senior and mid-level positions have reduced the ability of such ministries to play an effective role in defining their countries' energy policy and in monitoring energy developments in order to react quickly to market failures and other crises.[29] In view of the complex issues involved, governments should strive to establish and maintain a high level of policy-formulation capacity, without relying unduly on the regulatory agencies, whose task is to implement – not formulate – government policies.

## Clear separation of government's functions is best

As mentioned earlier, there is no catch-all, prescriptive approach to the functional organisation of governmental activities in the natural gas sector, but experience around the world has demonstrated the merits of clearly separating responsibilities for policy-making, regulation and dealing with state companies. This, in particular, improves the transparency and reduces the arbitrariness of decision-making, thereby reducing uncertainty, increasing predictability, lowering the cost of capital and increasing the attractiveness of investment.

## Setting a comprehensive approach to energy policy

In virtually all of its applications and uses, gas can be replaced by other sources of energy. Hence, to be effective, gas-sector policy must be part of a coherent and integrated energy-sector strategy. In emerging gas markets, particular attention should be given to the proper design of electricity markets, as gas-to-power projects are key to ensuring the financial viability of the whole gas chain, from wellhead through the pipeline to the gas-fired power plant and to the electricity consumer.

Co-ordination between the energy ministry and other government agencies is also important because energy policy has many interfaces with environmental, fiscal, social and industrial policy objectives, to mention just a few.

## Re-assessing the role of gas-for-power

Large pipeline projects need large off-takes, at least initially, to justify their commercial viability, and power generation is the only sector that can absorb large quantities of gas with a short lead time. Hence, in most cases, infrastructure development in gas-importing countries has been based on gas-for-power projects. In South America, the abundance of hydro-based electricity, a large part of which comes from written-off plants, puts a question mark on the role of gas-based power. Brazil is the most difficult case, as hydropower currently accounts for nearly 90% of electricity production and there is still substantial hydro potential. Under such conditions, it is not clear how gas-based power will compete with cheaper hydropower from already-amortised plants (and even from new hydropower plants). Where large-scale CCGT plants are not competitive, other options should be explored, such as building small and medium-scale gas-fired units closer to demand, developing gas-fired distributed generation and co-generation, and encouraging whenever possible the replacement of electricity appliances by gas-fuelled appliances (e.g. electric showers, air conditioning, some industrial applications).

---

29. This has been the case, for example, in Brazil, where the signs of an impending electricity crisis were visible long before it happened, but the government failed to properly appraise the problem and to take adequate preventive measures.

## Moving to competitive gas pricing and using taxation as a policy instrument

Competitive pricing of natural gas is key to the successful development of gas markets, as the price of gas relative to competing fuels will affect the volume of gas sold and the expected revenues from sales, ultimately determining the viability of any natural gas project. International experience has shown that pricing decisions all along the gas chain are best left to market players, with the state's role restricted to preventing anti-competitive behaviour and discrimination. In countries which do not have large and inexpensive gas reserves close to demand centres, a sound gas pricing system should start with the final consumers' willingness to pay, determined by the overall costs of using alternative fuels.

A basic requirement of this approach is that the prices of competing fuels should also be market-based and undistorted. The government can, however, improve the competitiveness of gas *vis-à-vis* alternative fuels for environmental or other policy objectives, for example by enforcing stricter emission standards, or eventually banning the use of more polluting fuels in certain applications or in certain areas.

Favourable taxation of gas *vis-à-vis* alternative fuels is another way of stimulating the supply of and demand for natural gas, which has been used by many countries, especially in the early stages of gas-industry development. Such use of taxation is especially justified if differences in taxes reflect differences in environmental externalities and costs, which are not accounted for in market-based prices.

## Adapting the institutional and regulatory framework to national circumstances

While the experiences of other countries may provide useful guidance, it is essential for each country to tailor its policies, institutional arrangements and regulatory framework to its specific circumstances, taking into account factors such as the level of development of gas markets; the extent to which the country is self-sufficient in gas or dependent on imports; the geographical distribution and number of gas suppliers; and its endowment with other energy resources, particularly for power generation.

Some of the institutional and regulatory choices made in North America, Western Europe and Argentina may only be of partial applicability to other South American countries, as those choices are adapted to countries with a well-developed gas industry and markets. It would be premature to adopt a similar approach in emerging gas markets. For example, in countries with a well-developed transport infrastructure and long-amortised costs, TPA and unbundling have shown to be effective in stimulating gas-to-gas competition, allowing for better use of the infrastructure and reduction of costs. However, in an emerging market, where significant investment in new pipelines is necessary and the market risk is high, enforcing vertical separation and TPA may deter investment in new pipelines and distribution networks.

## Reducing political, legal and regulatory risk

Because of their high up-front investment, long lead times and high specificity, gas projects are particularly sensitive to risk, as risk increases the financing cost. Some of the risks (i.e. commercial and technical risks) can be properly assessed and managed by the investor. However, governments have a special role to play in reducing specific risks for which they may be responsible, in particular risks of political, legal and regulatory nature. Investors will be reassured by the existence of clear, transparent and predictable legislation and regulation governing private investments in gas activities.

Laws and regulations may need to be changed over time, but investors should be protected from retroactive changes which may affect them detrimentally.

## Promoting cross-border co-operation and harmonisation

Given the distribution of resources and markets for gas in South America, cross-border trade will substantially increase in the years to come, both between neighbouring countries and for exportation in the form of LNG to the North American and European markets. These cross-border deals not only need clear policies for the upstream and downstream sectors in each of the countries involved, but also some degree of policy co-ordination between these countries. While policy and regulatory harmonisation across countries would facilitate cross-border projects, such harmonisation is likely to take a long time (as it is the case in Europe). In the meanwhile, governments should focus on creating the right conditions for investment in their own country and in developing bilateral or multilateral agreements, for example to avoid double taxation.

# PART II

COUNTRY PROFILES

# CHAPTER 5
# ARGENTINA

Argentina is a net hydrocarbon exporter and has a mature domestic gas market. A leader in the privatisation of state-owned utilities, Argentina implemented sweeping changes to its upstream and downstream gas industry in the early 1990s. Since then, growth in domestic gas demand has continued steadily. In recent years, the development of a regional gas pipeline network has enabled Argentina to increase gas exports to neighbouring countries and Argentine energy companies are looking increasingly to neighbouring Mercosur countries and beyond for investment opportunities. At the same time, many foreign companies have entered the Argentine oil and gas market, attracted by the favourable investment and tax environment and the prospects for regional expansion. The country's economic and financial difficulties are severely affecting the gas sector (as well as the rest of the energy sector). Companies, largely indebted in US dollars on the international markets, have seen their revenues in pesos devaluated by 400%. The gas distribution companies, already debilitated by three years of recession and affected by an increasing incidence of unpaid bills, are accumulating huge debts, which their international parent companies are unwilling to cover. The situation is no less difficult in the upstream sector. With the devaluation of the peso, the wellhead price – which is determined by the spot price in Buenos Aires – has been falling drastically, in some cases below production costs.

**Argentina at a glance, 2000**

|  | Argentina | Share in region |
|---|---|---|
| Surface area: | 2,780,400 km² | 16% |
| Population: | 37 million | 11% |
| Capital: | Buenos Aires | |
| Currency: | Peso Argentino (P$) | |
| GDP*: | US$426 billion | 18% |
| Total primary energy production: | 81 Mtoe | 14% |
| Primary energy consumption: | 62 Mtoe | 15% |
| Primary gas consumption: | 31 Mtoe | 35% |
|  | **Argentina** | **Regional average** |
| Per capita GDP*: | US$11,505 | US$6,821 |
| Per capita primary energy cons.: | 1.66 toe | 1.15 toe |
| Per capita electricity consumption: | 2,129 kWh | 1,708 kWh |

*   *Gross domestic product (GDP) expressed in 1995 prices and PPP.*

## BACKGROUND

**Economic development**

The second-largest South American country after Brazil, Argentina borders Chile, Bolivia, Paraguay, Brazil and Uruguay, and has a 5,000-kilometre coastline on the Atlantic Ocean. Its population of 37 million, the third-largest in the continent after Brazil and Colombia, is highly urbanised: 90% live in urban centres, compared to South America's average of 79%. Population growth has been slower than in the rest of the continent, and is expected to average 1% p.a. over the next 20 years.

Despite a long history of political instability and economic decline, Argentina remains the richest country in South America, as measured by GDP per capita. In 2000, per capita GDP (expressed in 1995 prices and adjusted for Purchasing Power Parities – PPP) stood at US$11,500, nearly double South America's average of US$6,800. The depreciation of the peso in 2002 narrowed this difference, but Argentina's GDP per capita remains the highest in South America.

This is, however, a pale reflection of a much brighter past. In the 1920s, Argentina was one of the world's wealthiest countries, with economic development and income levels comparable to Europe, Canada and the United States. From the end of the 1920s until the early 1990s, the Argentine economy deteriorated steadily. Indeed, from 1976 to 1989, Argentina had virtually no GDP growth, and its per capita income actually contracted by 1.1% annually. By 1990, Argentina's per capita income was substantially below that of the OECD countries, and had been overtaken by a number of rapidly-growing Asian nations.

**Economic reforms and privatisation**

In order to end a decade of foreign debt crisis, disarray in public finances and high inflation, the government of President Menem, that took office in 1989, launched into far-reaching structural reforms and a radical stabilisation plan, which included the adoption of a currency board. The Convertibility Law which pegged the peso to the dollar and backed each peso in circulation with a dollar in the Argentine Federal Reserve was introduced in April 1991, bringing an end to years of devastating hyperinflation, which had reached 5,000% per year. Argentina was one of the first South American countries to embark on IMF-backed austerity measures and to launch an aggressive and rapid privatisation programme. Almost all public enterprises, including power plants and electricity utilities, the telephone and airline companies, the television and radio stations, and the railways and highways, were sold to private investors. Shares in the state-owned oil-and-gas company, *Yacimientos Petrolíferos Fiscales* (YPF), and the national gas transmission and distribution company, *Gas del Estado* (GdE), were also sold off.

**The 2001 economic and financial crisis**

These efforts restored Argentina's credibility in the international financial community. They virtually eliminated inflation, attracted foreign and domestic investment and generated much higher rates of economic growth. As a result, Argentina was hailed as one of the most economically and politically stable countries in Latin America in the 1990s. However, the reforms failed to tackle the country's fiscal problems, which led to a rapid increase in public-sector debt. These structural trends were aggravated by external factors, including the 1998 Asian crisis, a weakening currency in neighbouring Brazil, a global economic slowdown and weak capital market, which led to a prolonged recession, starting at the end of 1998.

This culminated in a full-blown economic, financial and political crisis at the end of 2001, when, demoralised by three years of recession and deteriorating standards of living[1], Argentines took to the streets, pushing the already shaky government of President Fernando de la Rua to resign. Within days of taking office in January 2002, the new government abandoned the currency board and devalued the peso. After a brief attempt to sustain the new parity, the peso was left to float. By the end of 2002, the US dollar was worth around 3.5 Argentine pesos. GDP contracted by 4.4% in 2001, and the Economist Intelligence Unit (EIU) expects that it will fall by 11.3% in 2002.

## Foreign investment

Until the late 1980s, foreign investment in Argentina was hindered by a highly regulated economic environment. There were outright bans on foreign participation in some sectors, and discriminatory approval procedures for new operations. In 1989 and 1993, important amendments were made to the 1976 Foreign Investment Law. These amendments effectively put foreign investors on a level footing with domestic investors. Under the new legislation, foreigners are permitted to own up to 100% of virtually any business. Few investments now require government approval, and there are few reporting obligations. Approval is generally automatic and regulation, where required, is minimal.

The reforms of the past ten years have made Argentina one of the most liberalised countries among emerging markets throughout the world. The Argentine legal framework for foreign investment is simple and clear. Restrictions have been lifted on virtually all industries and sectors, except those of national strategic importance, such as defence air-traffic control. In the energy sector:

■ No direct investment is allowed in uranium mining or nuclear power generation, as a matter of national interest;

■ Non-mining activities in frontier areas need the permission of the National Superintendent of Frontiers, a section of the Ministry of Defence. Mining activities in such areas are, however, unrestricted.

## Regional integration

Argentina is a member of the *Mercado Común del Sur* (Mercosur), a customs union formed in 1991 by Argentina, Brazil, Paraguay and Uruguay. Bolivia and Chile subsequently became associate members. Trade within Mercosur is tariff-free. Expanded regional trade has helped improve Argentina's relations with its neighbours. For example, in the past, relations between Argentina and Chile were frequently tense, mainly because of territorial disputes over border areas. In 1978, war between the two countries was narrowly averted by international diplomacy. All border disputes ended with the 1999 ratification by the Chilean and Argentine parliaments of the Treaty of Peace and Friendship (which had been signed in 1991 by the two countries' presidents). The many cross-border gas pipeline and electricity transmission lines now linking the two countries are the best guarantee of continued good relations.

---

1.  According to World Bank indicators, 36% of Argentines lived below the poverty line in 2000.

Trade imbalances between Argentina and Brazil, exacerbated by their divergent exchange-rate policies (the Brazilian currency is allowed to float, whereas until recently Argentina ran a currency board and a fixed exchange rate), have been a source of tension between the two countries. In January 1999, the currency crisis in Brazil, Argentina's main export market, had an adverse affect on Argentine exports, prompting Argentina to impose import controls in July 1999. This sparked much indignation in Brazil, but the two countries quickly resolved the issue, averting any major fallout. However, the devaluation of the peso, the imposition of exchange and capital controls, and the deep recession will cause severe disruptions to Argentina's trade flows. Consequently, it is likely that relations with Brazil, which have been difficult since the devaluation of the Real in 1999, will be further strained in the near term.

## OVERVIEW OF THE ARGENTINE ENERGY SECTOR

**A very gas-intensive economy**

Argentina has a very gas-intensive economy. In 2000, natural gas accounted for half of total primary energy supply (TPES) and for more than one-third of final consumption. Oil is the other main source of energy, accounting for 38% of TPES (*Figure 5.1*). Combustible renewables (biomass) and waste, hydropower and nuclear power account for most of the remainder.[2] The share of gas in TPES has increased sharply since the 1970s, mostly displacing oil, especially heavy fuel oil.

Argentina is well-endowed with energy resources and is a net exporter of oil, natural gas and electricity. It exports crude oil and petroleum products mainly to Chile, Brazil, Uruguay and Paraguay, as well as small amounts of crude to the US Gulf Coast. Exports

**Figure 5.1** Primary energy supply in Argentina by fuel, 1975-2000

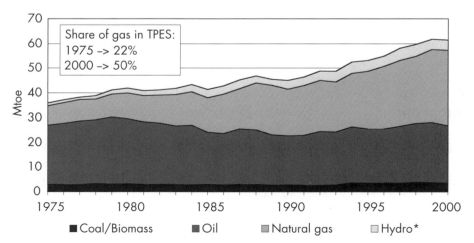

* *Includes nuclear and net electricity imports.*
*Source: IEA.*

---

2.  See Annex 1 for a detailed energy balance.

of natural gas to Chile, which began in 1997, are gradually increasing. Hydrocarbon exports accounted for 18% of the country's total export revenues for 2000. The country imports small amounts of electricity and coal.

As in the rest of South America, hydropower has an important role in Argentina's power generation. The country has 9,600 MW of hydropower plants, including the two large binational plants it shares with Paraguay (Yaciretá, 3,100 MW) and Uruguay (Salto Grande, 1,890 MW). Hydropower accounts for 40% of installed capacity and, depending on the amount of water available, generates 27% to 42% of total electricity output. Argentina has two nuclear power plants, one in Atucha, in the province of Buenos Aires, and another in Embalse, in the province of Córdoba, with a total installed capacity of 1,018 MW. A third reactor, the 600 MW Atucha II, has been under construction since 1981, but is not expected to become operational before 2007. The country also has significant solar and wind potential, and the government is putting in place measures to promote renewable energy particularly for isolated areas, but also for grid-connected projects.

**Figure 5.2** Electricity generation in Argentina, 1975-2000

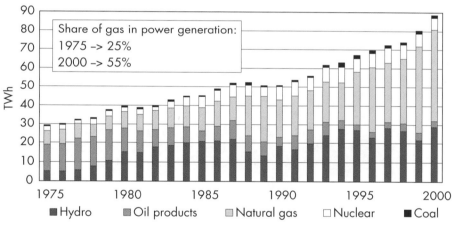

Source: IEA.

Primary energy supply grew 3.2% per year between 1990 and 2000, more slowly than GDP growth, which averaged 4.6% over the same period. The sectorial breakdown of final energy demand is close to the average for OECD countries, though transport accounts for a slightly higher proportion in Argentina. Energy demand in Argentina is highly concentrated. Buenos Aires, Argentina's capital and largest city, is the largest energy demand centre. The provinces of Buenos Aires, Córdoba and Santa Fé account for 80% of the country's energy demand.

Reflecting its high level of industrialisation and high per capita income, Argentina also has one of the highest per capita levels of primary energy use in South America: third after Trinidad & Tobago and Venezuela, whose large hydrocarbon and petrochemical industries account for a sizeable part of primary energy use. In 2000, Argentine per capita primary energy use was 1.66 tonnes of oil equivalent (toe), about half of the average for IEA Europe and 20% of the average for IEA North America.

Energy intensity, measured as TPES per unit of GDP adjusted for PPP, is lower than that of most other countries in South America and about 20% less than the average for OECD countries (see comparative tables in *Annex 3*). After rising steadily through the 1970s and early 1980s, energy intensity in Argentina has remained flat over the last ten years.

Due to the large share of natural gas in the Argentine energy mix, carbon intensity, measured as tonnes of energy-related $CO_2$ emissions per unit of GDP adjusted for PPP, is low compared to other oil and gas producing countries such as Venezuela and Trinidad & Tobago. But it is higher than in Brazil and other South American countries that rely substantially on hydropower. Per capita carbon emission levels in Argentina are the third-highest in South America, after Venezuela and Trinidad & Tobago.

**Energy-sector structure and institutions**

Argentina led the region's energy reform trend in the early 1990s. As part of an overall program of economic reforms and liberalisation, it restructured and privatised all its state-owned energy companies, including the oil and gas upstream company *Yacimientos Petrolíferos Fiscales* (YPF) and the gas transmission-and-distribution monopoly *Gas del Estado* (GdE). The Argentine energy sector is currently the most liberalised in the region, characterised by private companies in all segments of the industry.

The Ministry of Economy, Public Works and Services is the government body responsible for the energy sector. Within this ministry, energy issues are dealt with by the *Secretaría de Energía* (Secretariat of Energy), which is subdivided into two under-secretariats: one for electricity and one for hydrocarbons.

The National Regulatory Entity for Gas (*Enargas*) is the regulatory agency for the gas sector. The National Regulatory Entity for Electricity (*ENRE*) is its equivalent for the electricity sector. There is also a National Commission for Atomic Energy (*CNEA*).

Finally, the *Secretaría de Recursos Naturales y Medio Ambiente* (Secretariat of Natural Resources and Environment) is the federal body responsible for regulating environmental compliance by industry.

## NATURAL GAS RESERVES AND PRODUCTION

Argentina holds 11% of South American natural gas reserves (see *Table 1.1* in *Chapter 1*). Since 1970, the country's gas reserves have doubled. Most of the new reserves have been discovered as a result of oil exploration and are therefore concentrated in the same basins as oil production. There are 19 known sedimentary basins in the country, ten of which are located entirely onshore, three entirely offshore and six straddling the Atlantic coastline. *Map 6* at the end of the book shows the location of these basins.

Production is currently limited to five basins:

■ the Noroeste Basin in north-west Argentina;

■ the Neuquén and Cuyo Basins in central Argentina;

■ the Golfo San Jorge and Austral Basins in southern Argentina.

Productive basins account for around one-third of the total acreage of known sedimentary basins in Argentina. The Neuquén Basin accounts for almost 60% of current gas production and 50% of proven reserves (*Table 5.1*). To date there has been little offshore exploration. New discoveries in the offshore San Jorge basin have been disappointing. The government intends to offer incentives, in the form of lower royalty and tax rates, to companies engaged in oil and gas exploration in high-risk areas, both onshore and offshore.

**Table 5.1** Reserves and production of natural gas in Argentina's producing basins, 2001

| Basin | Proven reserves as of 1.01.2002 (bcm) | Probable reserves (bcm) | Gross production (mcm) | R/P ratio (years) |
|---|---|---|---|---|
| **Noroeste** | 161.7 | 64.9 | 7,827 | 21 |
| **Cuyo** | 0.5 | 0.1 | 76 | 7 |
| **Neuquén** | 377.9 | 95.2 | 25,892 | 15 |
| **Golfo San Jorge** | 47.4 | 27.7 | 3,186 | 15 |
| **Austral** | 176.0 | 117.5 | 8,932 | 20 |
| **Total** | **763.5** | **305.2** | **45,913** | **19***

\* *The R/P ratio of individual basins is calculated as proven reserves divided gross production, while the R/P ratio is calculated as usual.[3]*
*Source: Secretaría de Energía.*

Total proven reserves amounted to 764 bcm on 31 December 2001, the equivalent of 19 years of current production. However, a large part of remaining reserves is located in the Austral Basin, a little populated region some 4,000 kilometres south of Buenos Aires. Full use of the gas from the Austral and San Jorge basins will require a significant increase in the capacity of the southern gas transmission system.[4] The San Martin pipeline, which links Río Grande in the extreme south to Buenos Aires, is already working at full capacity (22.3 mcm/d in 2001). Apart from proven reserves, Argentina holds considerable potential for additional gas (and oil) reserves, as 14 sedimentary basins have not yet been explored.

The reform of the oil and gas sector and the privatisation programme implemented in the early 1990s brought substantial new investment into the Argentine gas market. Gross natural gas production accelerated, reaching 46 bcm in 2001 from 23 bcm in 1991. After declining in the 1980s and early 1990s, proven reserves started growing again, from 517 bcm in 1993 to 778 in 2000, but fell again to 764 bcm in 2001. However, much of the new reserves added between 1993 and 2000 came from workovers and re-appraisals of existing reserves, rather than from new discoveries. Low oil and

---

3.   The R/P ratio is normally calculated by dividing proven reserve by utilised production (gross production minus reinjection). According to Cedigaz data, gas reinjection in Argentina accounts for 7% of total gross gas production, hence the R/P ratios for individual basins in this table are slightly understimated.
4.   Or a commitment to LNG, which is unlikely at the moment because the reserves are not deemed sufficient.

gas prices and uncertainties about gas exports and planned pipelines reduced incentives for developers to explore additional gas fields. Other causes were the asset disposal following the merger of Repsol and YPF and delays in the introduction of new hydrocarbon legislation. A major upturn in oil and gas exploration occurred in 2001: the oil industry spent some US$3 billion that year, of which US$1.4 billion went to drilling operations. The main areas of interest remain the Noroeste, Neuquén and Austral basins. However, the economic and financial crisis has now stopped all investments in the country.

**Figure 5.3** Natural gas gross and marketed production, 1975-2001

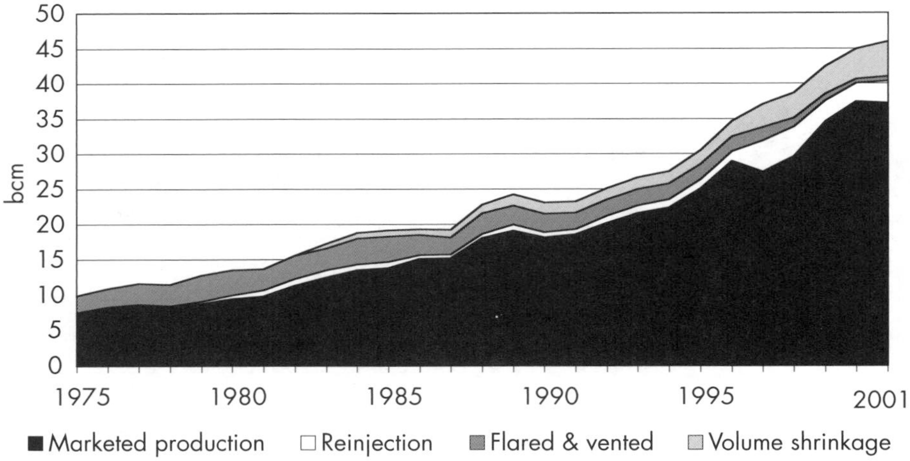

Source: Cedigaz.

**Main gas producers**

The Spanish-Argentine *Repsol-YPF* is the country's largest natural gas producer. *Yacimientos Petrolíferos Fiscales* (YPF) was the state-owned oil and gas company until 1992, when it was restructured, downsized and privatised. As of the year 2000, Repsol had acquired a 99% stake in YPF. Repsol-YPF is now Spain's largest company in terms of revenues, the largest private energy company in Latin America in terms of total assets and one of the world's ten largest oil companies on the basis of stock capitalisation and proven reserves. Repsol-YPF is engaged in all activities in the hydrocarbon sector. This includes the exploration, development and production of crude oil and natural gas; the transport of oil products, liquefied petroleum gas (LPG) and natural gas; oil refining; the production of a wide range of petrochemicals; and marketing of oil products and derivatives, petrochemicals, liquefied petroleum gas and natural gas.

In Argentina, Repsol-YPF controls 36% of total gas reserves and about 30% of natural gas production as an operator. However, through joint ventures with other companies and purchases from other producers, it actually supplies 51% of all gas in the wholesale market. In reality, these percentages are even higher because Repsol-YPF holds a 67.86% stake in the Argentine oil company Astra.

Repsol-YPF also has interests in distribution and transmission assets, though anti-monopoly regulations prevent its expansion in this area. In particular, through its 47% stake in the Spanish gas company GasNatural, Repsol-YPF has a 20% interest in Gas Natural Ban S.A., Argentina's second-largest gas distribution company. Repsol-YPF also has assets in power generation: it owns 45% of the 600-MW Pluspetrol Energy power station and 36% of the 845-MW Dock Sud power station.

The other two largest gas producers in Argentina are *Total Austral S.A.*, a fully-owned subsidiary of TotalFinaElf, and *Pluspetrol*, an Argentine company. Together, YPF-Repsol and these two companies control half of Argentina's gas reserves, and accounted for 58% of gas production in 2000. TotalFinaElf is increasing its presence in the Argentine gas sector. In late 2000, it acquired the Argentine and other Latin American assets of TransCanada Pipelines. TotalFinaElf also holds more than 60% of the capital of the 2,150-MW Central Puerto power station.

*Tables* 5.2, 5.3 and 5.4 show the reserve base and market share of Argentina's main gas producers.

**Table 5.2** Natural gas reserves by company in Argentina, as of 1 Jan. 2002

| | Proven reserves (mcm) | Probable reserves (mcm) | % of proven reserves |
|---|---|---|---|
| **Repsol-YPF** | 207,214 | 37,402 | 27% |
| **Total Austral** | 169,233 | 92,249 | 22% |
| **Pan American** | 98,726 | 54,824 | 13% |
| **Pluspetrol** | 85,080 | 2,674 | 11% |
| **Pecom Energía** | 63,160 | 33,922 | 8% |
| **Tecpetrol** | 58,838 | 36,531 | 8% |
| **Capex** | 21,180 | 2,114 | 3% |
| **Petrolera Santa Fé** | 18,083 | 11,870 | 2% |
| **Sipetrol** | 11,635 | 2,884 | 2% |
| **Chevron San Jorge** | 9,356 | 15,905 | 1% |
| **Pioneer** | 7,222 | 6,468 | 1% |
| Others | 13,799 | 8,393 | 2% |
| **Total** | **763,526** | **305,236** | **100%** |

Sources: Secretaría de Energía.

**Table 5.3** Natural gas production by operator in Argentina, 2001

| | mcm | % |
|---|---|---|
| **Repsol-YPF** | 13,520 | 29% |
| **Total Austral** | 8,173 | 18% |
| **Pluspetrol** | 5,391 | 12% |
| **Pan American** | 4,147 | 9% |
| **Pecom Energía** | 3,949 | 9% |
| **Tecpetrol** | 3,921 | 9% |
| **Petrolera Santa Fe** | 1,407 | 3% |
| **Chevron San Jorge** | 1,010 | 2% |
| **Capex** | 967 | 2% |
| Others | 3,426 | 7% |
| **Total** | **45,910** | **100%** |

Sources: Secretaría de Energía.

**Table 5.4** Share of natural gas supply by producer in Argentina, 2001-2002

| | |
|---|---|
| **YPF** | 51% |
| **Total-Pan American-Wintershall** | 9% |
| **Pecom Energía S.A.** | 7% |
| **Tecpetrol-Mobil-CGC** | 4% |
| **Propietarios de Sierra Chata** | 4% |
| **Pan American Energy** | 4% |
| **Pan American-Pionner-Coastal** | 3% |
| **Pluspetrol** | 3% |
| **Quintana-CGC-Sudelektra** | 3% |
| **Tecpetrol S.A.** | 2% |
| **Pluspetrol-Astra** | 2% |
| **YPF-Total-Pan American-Wintershall** | 2% |
| Others | 6% |
| **Total** | **100%** |

Sources: Enargas.

## NATURAL GAS DEMAND

**A mature gas market**

Argentina has a mature gas market, with the highest level of gas penetration in South America and one of the highest levels in the world. In 2000, gas accounted for 50% of the TPES, slightly more than in Venezuela and in the Netherlands (46%) and nearly as much as in Russia (52%). The share of gas in total final energy consumption[5] was 35% in 2000. For comparison, it was 38% in the Netherlands, 31% in Venezuela and 27% in Russia.

In 2000, total gas supply reached 36.5 bcm. At the end of that year, there were 5.9 million gas consumers in Argentina, of which 5.5 million were households. Between 1990 and 2000, primary gas demand grew at an average of 4.9% per year, compared with a 3.2% annual growth in total energy demand.

Since the start of the restructuring of the gas industry in 1992-93, consumption has risen in all sectors, but uses of gas in power generation and in transport have shown the fastest rates of growth. From 1992 to 2000, about 8 GW of gas-fired power plants came on stream, the number of service stations selling compressed natural gas (CNG) tripled and the number of vehicles fuelled with CNG reached 700,000. In the same period, 1.3 million new residential users and 60,000 new commercial and industrial users were connected to the grid.

**The power sector is the largest gas consumer**

The power generation sector is the country's largest gas user, with a 30% share. In Argentina, as in other countries, natural gas combined-cycle plants have become the most competitive power generation option and the primary technology choice for private developers. Much of the increase in electricity demand in the last decade has been met by additional gas-fired power plants. Gas-fired power plants now account for 56% of

---

5. See Glossary for definitions of primary energy supply and final energy consumption.

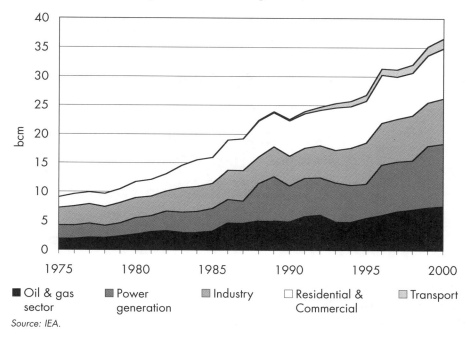

**Figure 5.4** Natural gas demand in Argentina, 1975-2000

Legend: ■ Oil & gas sector   ■ Power generation   ▨ Industry   ☐ Residential & Commercial   ▨ Transport

*Source: IEA.*

total capacity and, depending on the amount of hydropower available, generate between 35% and 57% of total electricity output. Use of gas by the power generation sector (including autoproducers) has doubled in the last decade, reaching 11 bcm in 2000 (see *Figure 5.2* above).

The industrial sector is the second-largest gas consumer, with a demand of 7.8 bcm in 2000 (21% of total gas supply). A substantial volume of gas (7.5 bcm in 2000) is used by the oil and gas sector itself.[6] The residential sector accounted for another 19% of gas supply in 2000, while the commercial and public sector and the transport sector accounted for 5% each. *Figure 5.5* shows the level of gas penetration in the different sectors.

**Major use of gas as a transport fuel**

Use of compressed natural gas (CNG) as a transport fuel is widespread in Argentina. Indeed, Argentina has the largest number of CNG vehicles in the world. At the end of 2000, there were 700,000 CNG vehicles and over 900 service stations selling CNG, in 17 provinces and 181 cities and towns. In Buenos Aires, almost all of the 50,000 taxis are CNG-fuelled, but public buses are still all diesel-fuelled. The government intends to increase CNG use for public transport in large cities. Since CNG is cheaper than gasoline and diesel, and the payback period for conversion to CNG is relatively short, CNG use is expected to continue growing.

---

6. For the purpose of this study, gas used in the oil and gas sector includes gas used in exploration and production, in refineries, and losses in transformation and transport. Use of gas for methanol production is included in the industrial sector (rather than in the transformation sector as in IEA statistics). Use of gas by the oil and gas sector does not include gas reinjected, flared or vented.

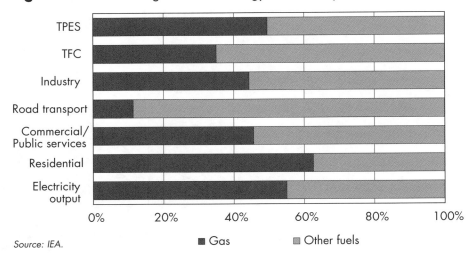

**Figure 5.5** Share of gas in total energy demand by sector, 2000

*Source: IEA.*  ■ Gas  ▨ Other fuels

**High seasonality of gas demand**

A characteristic of the Argentine gas system, which is not found in other countries of South America except Chile, is the high seasonality of gas demand due to the large use of gas for space heating in the residential sector. In 1997, residential demand was 4.7 times higher in July (the coldest month in the Southern Hemisphere) than in January. Demand in the commercial sector, and to a lesser extent in industry, is also sensitive to temperature. High seasonality, associated with a lack of storage capacity, means that when winter gas demand reaches system capacity, power generators, which hold the majority of interruptible gas contracts, must switch to oil and coal to conserve gas supplies for residential and commercial users. This problem is tempered by the fact that hydropower availability is also higher in the winter, alleviating winter demand for thermal generation.

**Future demand trends**

Given Argentina's already high level of gas use in all sectors, future growth in domestic gas demand is expected to be relatively low compared to developing gas markets such as those in neighbouring Chile and Brazil. According to Argentina's Secretariat of Energy[7], aggregate gas demand is projected to increase at an average annual rate of 3.9% between 2000 and 2010, a significantly lower rate than in the past decade (5.8%). As shown in *Table 5.5*, growth rates are expected to be highest for power generation and CNG. Per capita consumption of gas is expected to increase from 857 m³ in 2000 to 1,088 m³ in 2010. These projections will need to be revised downwards, as the economic and financial crisis has reduced energy demand growth.

**Gas transmission and distribution**

Argentina has a well-developed gas transmission and distribution network. At the end of 2001, the transmission network included 12,800 kilometres of high-pressure lines and 46 compressor stations, with a total throughput capacity of 121 mcm/d. The distribution network has 109,500 kilometres of low-pressure lines. The national transmission network includes five high-pressure pipeline systems, three of which bring gas from the Neuquén and Cuyo basins in the West, while the other two connect the Austral basin in the South and the Noroeste basin in the North. All five systems link into the Greater Buenos Aires market (see *Map 7* at the end of the book).

---

7. *Prospectiva 2000*, Secretaría de Energía, 2001.

**Table 5.5** Projected growth of Argentina's domestic gas demand

| Sector | 1990-95 | 1995-00 | 2000-05 | 2005-10 |
|---|---|---|---|---|
| Power plants | 2.1% | 12.4% | 6.6% | 3.3% |
| Industrial | 8.3% | 1.8% | 2.8% | 3.5% |
| Residential | 6.2% | 3.9% | 3.0% | 2.9% |
| Commercial | -2.0% | 2.1% | 3.0% | 2.9% |
| Transport | 35.8% | 10.7% | 5.7% | 5.5% |
| **Total** | **5.8%** | **5.8%** | **4.4%** | **3.4%** |

| Per capita demand | 1995 | 2000 | 2005 | 2010 |
|---|---|---|---|---|
| Cubic metres | 671 | 857 | 972 | 1,088 |

*Source: Secretaría de Energía.*

Gas transmission is carried out by two large companies. *Transportadora del Norte (TGN)* owns 5,400 kilometres of pipelines, with a total capacity of 52.3 mcm/d, covering the northern and central regions of Argentina. The TGN network receives gas from the Neuquén and Noroeste basins and connects with the Bolivian gas network in Campo Durán (Salta Province). It serves the main cities in northern and central Argentina, the industrial district of Rosario-Buenos Aires and the agglomeration of Greater Buenos Aires. Three of the cross-border pipelines linking Argentina to Chile (NorAndino, Atacama and GasAndes) are connected to the TGN network. TGN is jointly owned by *Gasinvest S.A.* (70%) and US-based *CMS Energy* (30%). Gasinvest S.A. is a consortium of French TotalFinaElf (which bought TransCanada Pipelines' interest in 2000), Argentine Techint and Compañía General de Combustibles S.A., and Malaysian Petronas.

*Transportadora del Sur (TGS)* serves the centre and south of Argentina. It operates the most extensive gas pipeline system in Latin America, with more than 7,400 kilometres of pipelines and a transportation capacity of 64.2 mcm/d. The TGS system transports gas from the Austral, San Jorge and Neuquén basins and serves the Greater Buenos Aires area as well as the mostly-rural provinces of western and southern Argentina. *Compañía de Inversiones de Energía S.A.* (CIESA) holds approximately 70% of TGS's equity, while local and foreign investors in the New York and Buenos Aires stock exchanges hold the remaining 30%. CIESA is owned by Argentina's Pecom Energía S.A. (a wholly-owned subsidiary of Perez Companc S.A.) and US-based Enron Corp., each with a 50% stake.

There are nine distribution companies, each with exclusive rights over a given geographical area (see *Map 7*). *Metrogas S.A.*, which covers most of the Buenos Aires metropolitan area, is Argentina's largest natural gas distribution company, in terms of both number of clients and volume of gas deliveries. It supplies over 1.8 million customers in Buenos Aires, accounting for a quarter of all gas sold by distribution companies. Metrogas's licensed area includes the federal capital and the southern and eastern areas of Greater Buenos Aires. The distribution company has natural gas purchase contracts with 15 producers, principally in southern and western Argentina. Metrogas's largest supplier, Repsol-YPF, provides about 45% of the company's total gas requirements. Gas Argentino S.A. is the majority shareholder in Metrogas, controlling 70% of the company's shares. British BG International, the technical operator of Metrogas, has a 55% share in Gas Argentino. The other main partners are

Repsol-YPF's subsidiary Astra S.A., with 27%, and Argentine Private Development Co (APDC), with 19%. The remaining 30% of Metrogas's equity is held by former employees of the former state-owned gas company Gas del Estado (10%) and by private investors in the New York and Buenos Aires stock markets.

**Table 5.6** Natural gas distribution companies in Argentina, 2001

| Company | Number of of customers ('000) | Gas sales (mcm/d) | Length of network (km) | Investments 1993-2001 (US$ million) | Operator |
|---|---|---|---|---|---|
| Metrogas | 1,936.5 | 18.8 | 15,678 | 524.3 | BG International |
| Gas Natural Ban | 1,231.4 | 9.0 | 20,255 | 443.1 | Gas Natural |
| Camuzzi Pampeana | 425.6 | 7.8 | 21,545 | 342.8 | Camuzzi |
| Litoral Gas | 904.1 | 4.4 | 9,170 | 187.7 | Tractebel |
| Camuzzi Sur | 414.6 | 7.3 | 13,094 | 86.2 | Camuzzi |
| Gas del Centro | 417.1 | 4.8 | 12,088 | 112.6 | Società Italiana per il Gas |
| Cuyana Gas | 355.1 | 15.4 | 8,697 | 124.4 | Società Italiana per il Gas |
| Gasnor | 322.3 | 11.4 | 7,029 | 100.0 | Gasco |
| Gasnea | 12.1 | 0.3 | 1,977 | 8.2 | Gaseba |
| **Total** | **6,018.8** | **79.0** | **109,533** | **1,929.3** | |

Source: Enargas.

The two next largest distribution companies are *Gas Natural Ban S.A.* and *Camuzzi Pampeana S.A.* Gas Natural Ban is operated by Spanish Gas Natural, itself partly owned by Repsol-YPF (47%). Its concession area covers the northern and eastern parts of Buenos Aires and Greater Buenos Aires. Camuzzi Pampeana is operated by the Italian distributor Camuzzi, and supplies southern Buenos Aires, a considerable part of the Buenos Aires Province and the northernmost area of La Pampa up to the Colorado River.

Some of the private companies that operate and own equity in Argentina's other gas distribution companies are: Belgium's Tractebel (operator of *Litoral Gas S.A.*); Gaseba, a company of the Gaz de France Group (*Gasnea S.A.*); US Gasco (*Gasnor S.A.*); Italy's Società Italiana per il Gas p.a. (*Gas del Centro S.A.* and *Gas Cuyana S.A.*); and Italy's Camuzzi (*Camuzzi Gas del Sur S.A.*).

There has been significant expansion of the transmission and distribution networks since the restructuring and privatisation of the previously state-owned downstream gas company Gas del Estado. Transmission capacity rose by 60% between 1992 and 2000, while the length of the distribution network grew by 58% during the same period. Total investments for the 1993-2000 period reached US$1,962 million for transmission and US$1,764 million for distribution. Rising gas demand, the need to reduce pre-existing transmission bottlenecks, and government incentives, all played a role in promoting network expansion. For example, the government imposed in the process of privatisation a certain level of "obligatory initial investment" on newly-privatised transmission and distribution companies. Despite the expansion, there are still capacity constraints and bottlenecks, especially on the San Martin pipeline from Ushuaia to Buenos Aires, which is operating at full capacity.

There are currently no underground storage facilities in Argentina, though a number of projects to develop aquifers and depleted gas fields for storage use are being studied. BAN, the distribution company in Northern Buenos Aires, operates the country's only LNG peak-shaving plant, located at General Rodriguez near the capital. A second plant close to Buenos Aires is planned.

## NATURAL GAS TRADE

### Natural gas imports

Argentina imported gas from Bolivia from 1972 to 1999. Annual imports increased gradually from 4.2 mcm/d to 6 mcm/d. Bolivian imports provided 20% of Argentine gas demand in 1972, but only 5% in 1998. On the other hand, exports to Argentina accounted for more than 90% of Bolivian gas production in the 1970s, decreasing to 60-70% in the last decade. The long-term contract with Bolivia expired in 1997, but the two governments agreed to renew it to 1999, when Bolivian exports to Brazil started. Currently, Argentine Pluspetrol imports small quantities of gas from its fields of Bermejo y Madrejones in Bolivia, close to the Argentine border, to fuel a power plant and a refinery. Large-scale imports from Bolivia are unlikely to resume in the foreseeable future.

### Natural gas exports

Argentina currently exports gas to Chile, Brazil and Uruguay. There is a greater potential for growth in gas exports than in the domestic market. Numerous cross-border pipelines were built in the last five years (see section on cross-border pipelines). Exports to Chile started in 1997, rapidly building up to 5.3 bcm in 2001. Exports to Brazil began in 2000 and reached 740 mcm in 2001. In addition, Argentina exports smaller volumes to Uruguay. Total exports amounted to 6.1 bcm in 2001, earning a total of US$309 million.

*Table 5.7* summarises the recent evolution of and official projections for gas exports to 2010. Future exports could be even higher than these figures. Projections for future gas demand in Chile suggest that required gas imports from Argentina to Chile's central region alone, via the GasAndes and Pacífico pipelines, may in themselves exceed all of Argentina's officially projected exports to Chile. Several new cross-border pipelines are under construction to bring Argentine gas to Brazil's energy-hungry Southern and Southeastern regions. While Bolivian gas presently has preferential access to the Brazilian market, the Brazilian government favours the development of other cross-border pipeline projects to diversify sources of supply and bring down prices.

There were so many authorisation requests for gas export projects that a few years ago the Argentine government became concerned about the country's ability to meet increasing export commitments while continuing to supply a growing internal demand. It was argued that proven reserves should not be committed to additional export projects, since even at the current level of demand (including the amounts already committed to Chile) the reserve-to-production ratio is only about 20 years. However, the rapid growth of proven reserves in nearby Bolivia has somewhat alleviated such concerns.

**Table 5.7** Actual and projected exports of Argentine natural gas to neighbouring countries, 1997-2010 (mcm)

| Country | Pipeline | 1997 | 1998 | 1999 | 2000 | 2001 | 2005 | 2010 |
|---|---|---|---|---|---|---|---|---|
| **Brazil** | Paraná-Uruguayana | | | 171 | 740 | 899 | 950 | |
| | Cuiabá | | | | | | 347 | 381 |
| | **Total Brazil** | | | | **171** | **740** | **1,246** | **1,331** |
| **Chile** | GasAtacama | | | 202 | 533 | 811 | 1,205 | 1,299 |
| | GasAndes | 114 | 1,176 | 1,970 | 1,969 | 2,153 | 2,424 | 2,546 |
| | NorAndino | | | 9 | 213 | 598 | 1,495 | 1,767 |
| | Pacífico | | | 1 | 107 | 191 | 962 | 1,030 |
| | Patagónico | | | 101 | 287 | 326 | 378 | 382 |
| | Tierra del Fuego | 554 | 740 | 703 | 723 | 595 | 691 | 697 |
| | El Cóndor-Posesión | | | 317 | 582 | 600 | 696 | 703 |
| | **Total Chile** | **669** | **1,916** | **3,304** | **4,413** | **5,275** | **7,851** | **8,423** |
| **Uruguay** | Gasod. del Litoral | | 2 | 23 | 37 | 36 | 66 | 66 |
| | Cruz del Sur | | | | | | 1,040 | 1,588 |
| | **Total Uruguay** | | **2** | **23** | **37** | **36** | **1,106** | **1,654** |
| **Total exports** | | **669** | **1,918** | **3,327** | **4,621** | **6,051** | **10,203** | **11,408** |

*Source: Secretaría de Energía.*

Until recently the government also kept tight control over natural gas exports. Companies intending to export gas were required to apply to the Secretariat for Energy and Mining for an export licence. This was to ensure that exports would not jeopardise long-term domestic supply. Because of the significant increase of gas reserves in the region, notably in neighbouring Bolivia, export requests are now approved automatically.

## Cross-border gas pipelines

### Argentina-Bolivia

*Map 8* at the end of the book shows Argentina's cross-border gas links with neighbouring countries. The pipeline connecting Argentina and Bolivia was the first cross-border gas project in Latin America. Initiated in 1970, it connected the Argentine northern transportation network in Campo Duran (Salta, Argentina) with Río Grande in the gas-rich region of Santa Cruz de la Sierra (Bolivia), crossing the border at Yacuiba/Pocitos. The Argentine portion is operated by TGN, while the 441-km, 24-inch, 6-mcm/d Bolivian portion, also known as the *Yacimientos-Bolivian Gulf (Yabog)* pipeline, is operated by Transredes, a joint venture between US Enron (25%), Shell (25%), Bolivian pension funds (34%) and other investors (16%).

During the early stages of the construction on the Bolivia-Brazil (Gasbol) pipeline, when Bolivian proven reserves were still insufficient to cover the 20-year gas-sale agreement with Brazil, it was suggested that, as a fall back solution, the gas flow on the Yabog could be reversed to allow Argentina to export gas from the Noroeste Basin to Brazil. Now that Bolivian reserves have increased sharply, Argentine gas exports via Bolivia are less likely. The flow on the Yabog has indeed been reversed to bring gas from the rich southern Bolivian fields to the connection with the Gasbol pipeline. However, the Yabog pipeline is being expanded and there are also plans to build a new pipeline alongside it. It is not excluded that gas from the nearby Argentine fields will flow through this new pipeline to Brazil. Indeed, the Cuiabá power station in Brazil has already contracted gas from Repsol-YPF in Argentina.[8]

Currently, Argentina's Pluspetrol imports small quantities of gas from the Bolivian fields of Bermejo y Madrejones, close to the Argentine border, through small company pipelines linking Bermejo (Tarija, Bolivia) to Ramos (Salta, Argentina) and Madrejones (Tarija, Bolivia) to Campo Durán (Salta, Argentina).

**Argentina-Chile**

Argentina started exporting natural gas to Chile in 1996, and Chile is currently the largest market for Argentine gas. Seven pipelines link Argentina to Chile, all built between 1996 and 1999:

- three small pipelines link Southern Argentina's Austral basin to Punta Arenas (*Tierra del Fuego, El Cóndor-Posesión* and *Patagónico*);

- two pipelines supply Chile's populous central and southern regions (*GasAndes* and *Gasoducto del Pacífico*); and

- two others bring gas to the industries in the north of the country (*GasAtacama* and *NorAndino*).

The first cross-border pipeline connecting Argentina to Chile was the 14-inch, 83-km *Tierra del Fuego* pipeline, also called the *Methanex PA*. With a capacity of 2 mcm/d, it was built to meet the increased gas demand resulting from the expansion of a methanol-producing plant in the extreme south of Chile. It came on stream in 1996. The other two southern pipelines, the 9-km *El Condor-Posesión* and the 33-km *Patagónico* (also called *Methanex YPF* and *Methanex SIP*), came into operation in 1999 for the same purpose and have throughput capacities of 2 and 2.8 mcm/d, respectively.

The 24-inch, 463-km *GasAndes* pipeline, which extends from La Mora, Mendoza, in west-central Argentina to San Bernardo on the outskirts of the Chilean capital, Santiago, began operations in August 1997. It cost US$325 million. With a throughput capacity of 9 mcm/d, it supplies gas from Argentina's Neuquén Basin mainly to Metrogas, the gas distribution company for Santiago, and to a 379-MW power plant in Nueva Renca, Chile. GasAndes began as a consortium between TransCanada (with a 46.5% share), the Argentine Compañía General de Combustible (17.5%), the Chilean power generator GENER (13%) and Santiago gas distributor Metrogas (13%). In July 1998, TotalFinaElf acquired a 10% stake in the pipeline, and following its purchase in May 2000 of TransCanada's South American assets, TotalFinaElf's stake in GasAndes rose to 56.5%.

The 638-km *Gasoducto del Pacífico* (Pacific pipeline) transports gas from the Neuquén area of Argentina to the Bio-Bio region in the southern part of central Chile. Commercial operations began in November 1999. The cost of the pipeline was estimated at US$350 million. It has diameters of 10, 12, 20 and 24 inches and a throughput capacity

---

8. This power station is connected to the Gasbol pipeline through an offshoot starting in San Miguel, Bolivia.

of 9.7 mcm/d. The gas comes from the Neuquén Basin, and supplies industrial clients (cellulose, paper, cement, steel and glass factories, and fish and sugar processing plants). The pipeline also supplies commercial and residential users in Concepción, the second-largest city in Chile. Gasoducto del Pacífico began as a joint venture led by TransCanada, US-based El Paso Energy, Repsol-YPF, Chile's Gasco and the Chilean state-owned oil company Enap. TransCanada sold its shares in Gasoducto del Pacífico to TotalFinaElf in May 2000.

The 24-inch, 941-km, 8.5-mcm/d *GasAtacama* pipeline connects the Argentine network in Corneja, Salta, to the city of Mejillones on the northern coast of Chile. It was completed in July 1999 at an estimated cost of US$350 million. It supplies gas from Argentina's gas fields in the Noroeste Basin to power plants and industrial clients in the north of Chile. In addition to the natural gas pipeline, the Gas Atacama project includes two gas-fired generating plants on the Chilean side. The GasAtacama consortium includes US-based CMS Energy Corporation (40%), Chilean power company Endesa (40%) and two Argentine companies, Pluspetrol (16%) and Repsol-YPF (4%).

The 1,066-km, 12-, 16- and 20-inch *NorAndino* pipeline also links Argentina's Noroeste Basin in Salta province to Mejillones in Chile's northernmost region. With a transportation capacity of 7.1 mcm/d, it supplies industrial clients in Chile's Atacama desert region in the north, including the Tocopilla power plant and the world's largest copper-mining industry. It came into operation in November 1999. The NorAndino is a joint venture between Belgium's Tractebel (the operator, originally with a 66.6% stake) and US-based Southern Electric (owner of Chilean power generator Edelnor, which originally owned 33.3%). In July 1999, Tractebel raised its stake in the project to 68.2% whilst Southern Electric retained 31.8% through a special-purpose company called Energía del Pacífico.

**Argentina-Uruguay**

Argentina exports gas to Uruguay through the 26-km, 0.7-mcm/d *Gasoducto del Litoral*. This pipeline, which began operations in late 1998, links Colón in the Argentine province of Entre Ríos to Paysandú in Uruguay. A new branch is planned to be built later to supply a proposed 360-MW thermal power plant at Casablanca, Uruguay.

Construction of a new pipeline linking Buenos Aires with Colonia and Montevideo, the Uruguayan capital, started in March 2001. The 208-km, 5-mcm/d *Cruz del Sur* pipeline will cost an estimated US$170 million. Gas deliveries started in late 2002. The gas comes from the Neuquén Basin. While the first target of the project is the Uruguayan market, Cruz del Sur is the first segment of a bigger project to link Argentina to southern Brazil. The Cruz del Sur consortium includes Britain's BG with a 40% stake; Pan American Energy (a joint venture between UK-based BP and Argentine Bridas Energy) with 30%; Germany's Wintershall with 10%; and the Uruguayan state oil company Ancap with 20%. Pan American and Wintershall have already signed a 15-year agreement to supply the Uruguayan state electric company UTE with 1.75 mcm/d of gas from their Argentine reserves through Cruz del Sur.

**Argentina-Brazil**

Since July 2000, Argentina has exported gas to Brazil through a 440-km 24-inch pipeline connecting the TGN network at Aldea Brasilera, Paraná, in the centre-east of Argentina, to Uruguayana, a Brazilian town on the border with Argentina. The

*Paraná-Uruguayana* pipeline supplies a 600-MW thermoelectric power station at Uruguayana, owned by the US company AES. The Argentine portion of the pipeline is operated by Transportadora de Gas de Mercosur (TGM), a joint venture of Techint (Argentina), TotalFinaElf (France), Petronas (Malaysia), Compañia General de Combustibles (Argentina) and CMS Energy (US).

The short 25-km stretch in Brazilian territory is operated by Transportadora Sul Brasileira de Gas (TSB), a joint venture between Gaspetro, Petrobras' wholly owned natural gas subsidiary (25%); French TotalFinaElf (25%); Brazilian Ipiranga (20%), Spanish Repsol-YPF (15%), and Argentine Techint (15%). This is the first spur of a more ambitious project to supply Argentine gas to Porto Alegre, in Brazil's southern State of Rio Grande do Sul, where it would compete with Bolivian gas arriving through the Gasbol pipeline.

When concluded, the TGM-TSB pipeline will extend 1,055 kilometres and eventually have a capacity of 12-15 mcm/d. The pipeline was expected to be operational by mid-2002, but construction has been delayed for nearly 2 years because of uncertainties both on the demand side in Brazil and on the supply side in Argentina. In Brazil, difficulties connected to the definition of new electricity and gas sector regulations have delayed the construction of two gas-fired power stations near Porto Alegre, which will take some 5 mcm/d of the pipeline capacity. In Argentina, a capacity expansion of the TGN network is needed to supply the new line, but the crisis has put a halt on all capacity expansion plans.

**Planned pipelines**

Several other pipelines to transport Argentine gas to southern and southeastern Brazil's industrial and population centres have been proposed or are at various stages of development. The most advanced is an extension of the *Cruz del Sur* pipeline (see above). Starting in Colonia, Uruguay, this second spur will extend 415 kilometres in Uruguay and 505 kilometres in Brazil to Porto Alegre. This US$400 million project aims at supplying 12 mcm/d to the very same market aimed at by TGM-TSB pipeline.

Other Argentine-Brazil cross-border projects that had been discussed in past years are the *Mercosur* and the *Trans-Iguaçu* pipelines. The Mercosur project envisages linking Salta in north-western Argentina to Asunción, Paraguay, then to Curitiba, Brazil, and eventually to São Paulo. The pipeline would be 3,100 kilometres long and have a capacity of 25 mcm/d; however there were doubts about the adequacy of reserves in Argentina's Noroeste Basin to support the expected investment of US$1.8 billion. The *Trans-Iguaçu* project would cross from Argentina's Noroeste Basin into southern Brazil, with a capacity of 35 mcm/d.

Even with very optimistic expectations of gas-demand growth in southern Brazil, only one or two of the above pipelines will be needed in the medium term. Southern Brazil has substantial hydropower and also cheap local coal. The ultimate objective, however, is to supply the much bigger markets of São Paulo and Rio de Janeiro, where Argentine gas could successfully compete with Bolivian gas.[9] However, expansion of the *Gasbol* pipeline is now under consideration, and how much gas will eventually

---

9.   This would require the flow on the southern leg of the Gasbol (São Paulo-Porto Alegre) to be reversed.

flow from Argentina to Brazil will depend in large part on "who gets there first", as well as on future gas finds in Brazil's prolific basins offshore the States of Rio de Janeiro and São Paulo.

The Gasin project

In December 2001, the Brazilian government announced plans for a major gas pipeline project linking Brazil with Bolivia, Argentina and Paraguay. The so-called Gas Integration (*Gasin*) pipeline would extend 5,250 km in the four countries and cost an estimated US$5 billion. The pipeline would start near the gas fields in Southern Bolivia and run through gas-rich northern Argentina, where a spur would cross the border to Asunción, the capital of Paraguay. The pipeline would then enter Brazil at the western border of the Santa Catarina State and run 3,450 km inside Brazil, east-wise and then north-wise all the way to the federal capital Brasilia. Petrobras and Italy's Snam, a company of the ENI group, have already signed an agreement to begin the economic and technical feasibility studies, together with the Argentine Tiete-Parana Development Agency (ADTP) which first drew up the basic concept for the project. According to initial plans, the pipeline could be ready to start operation in 2005.

## GAS-SECTOR REFORM AND PRIVATISATION

During the 1990s, Argentina's gas sector underwent profound structural and regulatory changes to introduce competition and attract private investment. Gas sector reforms started in 1989, as part of an overall program of economic reforms and liberalisation, formalised by the State Reform Bill (Law No. 23.696/89), which, *inter alia*, mandated the restructuring of the gas sector. By early 1991, the upstream oil and gas sector had been largely deregulated, with most portions already open to private and foreign participation. In 1992, the Privatisation Law allowed privatisation of the state-owned *Yacimientos Petrolíferos Fiscales* (YPF). Also in 1992, the Natural Gas Act (Law No. 24.076/92) introduced sweeping changes to downstream gas-sector policy and regulation, providing for the vertical and horizontal unbundling and privatisation of the gas transmission-and-distribution monopoly *Gas del Estado* (GdE).

Until 1989, YPF had exclusive rights over oil and gas exploration and production in Argentina. These rights were based on the 1967 Hydrocarbons Law and subsequent decrees. For many years YPF's hydrocarbons production stagnated, and the company gained a reputation for inefficiency. Technologically, it lagged behind other large state-owned oil companies in the region. Several Argentine governments used the company as a "cash-cow" to resolve their budgetary problems. As a result, YPF suffered large financial losses, totalling around US$6 billion in the period 1981-89. Acute lack of capital investments resulted in steadily shrinking hydrocarbon reserves.

GdE had a monopoly on natural-gas transport and distribution in the whole country. Its long-standing dominance resulted in high costs and insufficient investments in transmission and distribution networks. Before the reform, capacity constraints and bottlenecks resulted in frequent supply interruptions during peak periods.

## Opening of the upstream and privatisation of YPF

Starting in 1989, several decrees abolished YPF's exclusive rights to exploration and production, removed wellhead price controls, eliminated restrictions on the oil and gas trade, and removed taxes and tariffs on exports. The objective of these decrees was to encourage investment in gas exploration and production and to foster competition in wholesale gas supply.

Preparations for the privatisation of YPF began with the downsizing of the company, the rationalisation of its operations and improvement in its general management. YPF was required to sell some of its oil and gas reserves (Decree 1055/89). Another decree (1212/89) allowed the renegotiation of sales contracts between private operators and YPF. From 1990 to 1995, YPF's work force fell from 51,000 to 5,800, and a third of the company's oil and gas reserves were sold off.

Privatisation of a substantially-reduced YPF was authorised by the Privatisation Bill in 1992. The sale was carried out in two phases. First, some of the company's assets were sold to the private sector for over US$2 billion. Then, stakes in the company YPF, which was retaining the core business, were sold through an International Public Offering (IPO) in 1993. When the IPO was completed, approximately 60% of the company was owned by private shareholders, with the federal government retaining 20%, the provinces 11% and employees 10%. By 1998, the ownership structure had changed dramatically: the employees had sold their shares in 1997, the provinces had reduced their stake to less than 5%, and the federal government was talking of selling its 20%. This paved the way to a take-over by Spanish Repsol, and in 1999 YPF became Repsol-YPF. By 2000, Repsol owned 99% of Repsol-YPF's shares.

## Concerns about market concentration

Despite the diversification of players achieved through the divestment of a third of YPF's assets and the exploration and production licences awarded since 1990, Repsol-YPF remains the dominant producer of gas in Argentina. In 2001, Repsol-YPF accounted for 29% of natural gas production, but wholesale activity is even more concentrated because, through its participation in joint ventures with other companies and purchases from other producers, Repsol-YPF actually supplies 51% of all gas in the wholesale market. By comparison, the second-largest producer, TotalFinaElf, in association with the US companies Pan American and Wintershall, only accounts for 9% (*Table 5.4* above). Repsol-YPF thus may be able to play the role of price-setter. It can impose prices, price formulae and other contractual terms on buyers, though political considerations have so far deterred the company from exercising its market power fully. Repsol-YPF is also the dominant player in the domestic oil-product market, where it exercises considerable influence over the prices of fuels that compete with natural gas in Argentina.

The government is aware of the problem, but has limited room for action under the existing legal and institutional framework. Repsol-YPF is now in private hands, and Argentine competition law, like that of European countries, puts the burden of proof on third parties to prove abuse of a dominant position, rather than outlawing dominance, as in the United States.[10] At the time of the Repsol-YPF merger, the Argentine authorities stipulated that the newly-formed company sell some of its refining and

---

10. For a detailed discussion of the problem and the options available to solve it, see Chapter VII, *Regulatory Reform in Argentina's Natural Gas Sector*, IEA, 1999.

downstream oil assets in Argentina, in order to reduce its dominant position. To meet this condition, Repsol-YPF entered into several asset-swap agreements with other oil companies in the region.[11] In addition, the government is putting pressure on Repsol-YPF to reduce its dominant position in the wholesale gas market. As a result, the company's share in total domestic gas supply decreased from 62% in 1997 to 44% in 1999, although it increased again to 51% in 2000 and 2001.

**Awaiting a new hydrocarbon bill**

A new hydrocarbon law has been awaiting consideration in the Argentine Congress since 1995, stalled by successive cabinet crises. The deadline for discussion passed, but the bill was resubmitted in 1998. As of mid-2002, the bill had not yet been submitted to Congress, delayed in 2000 and 2001 by other important bills and more recently by Argentina's economic and political crisis. Its main features, which are broadly supported by both domestic and foreign oil-and-gas companies, are the following:

■ the transfer of jurisdiction and administration of the upstream oil-and-gas industry from the federal government to the provinces, which would be entitled to set and collect taxes and royalties (up to a ceiling of 12%) and to license exploration acreage not already awarded by the federal government;[12]

■ the creation of a federal hydrocarbons regulatory agency, to be known as Ente Federal de Hydrocarburos (EFH), to regulate the industry at federal level;

■ incentives – such as increasing the length of exploration periods, reducing relinquishment obligations, and reductions in taxes and royalties – for exploration in frontier areas outside the five basins currently in production;

■ improved rents for landowners at oil and gas field sites;

■ the establishment of environmental and consumer-protection regulations;

■ the creation of a strategic oil reserve;

■ the introduction of ceilings to oil companies' share of the domestic oil-products market.

**New downstream set-up**

The 1992 Natural Gas Bill (Law No. 24.076/92), in conjunction with several decrees (notably Decree No. 11.739/92) and asset-transfer agreements, established a new institutional and regulatory framework for the downstream gas sector. The key elements of the reform were:

■ a clear separation of production, transmission and distribution activities, with the prohibition for companies involved in one segment to hold a controlling stake in companies operating in another segment;

---

11. One of these swap agreements is being finalised with Brazilian Petrobras, which is also under pressure to reduce its dominance in the Brazilian domestic market. Under the proposed deal, valued at over US$1 billion, Repsol-YPF will relinquish to Petrobras 734 service stations in Argentina and a controlling stake in a 30.5-kb/d refinery, in exchange for a 30% stake in one of Petrobras's refineries and a 10% stake in the promising Albacora Leste offshore field in Brazil. Another asset-swap was concluded in December 2000 with Chilean state-owned oil company Enap's international subsidiary Sipetrol.

12. Many legislators are concerned that this may push provinces to compete for foreign investors by offering lower royalty rates. This could exacerbate the differences between richer and poorer provinces.

■ the vertical and horizontal unbundling of the gas transmission and distribution assets of the state-owned monopoly GdE, and the privatisation of the resulting companies;

■ non-discriminatory third party access to all the country's transmission and distribution networks,[13] with regulated transmission and distribution tariffs set by the regulator and reviewed every five years;

■ liberalisation of wholesale gas prices and definition of "eligible customers", i.e. customers using more than a certain threshold amount of gas, who may, therefore, participate directly in the wholesale market, bypassing the distribution companies;

■ the creation of a spot market for gas;

■ the possibility to resell transportation capacity on a secondary market;

■ definition of the role of the trader or retailer;

■ the creation of an independent regulatory authority for gas, the Ente Regulador del Gas (Enargas), responsible for overseeing the regulated segment of the industry (transmission and distribution), protecting consumer interests and promoting competition in gas supply.

## Unbundling and privatisation of GdE

Before privatisation, GdE was reorganised on a broadly geographical basis and split into two transmission companies, *Transportadora Gas del Norte* (TGN) and *Transportadora Gas del Sur* (TGS), and eight medium/low-pressure distribution companies (a ninth distributor, covering the northeast of Argentina was subsequently established and licensed in 1997). The government then sold controlling stakes in these ten companies, requiring that the buying consortia include an international operator as well as a local partner. This led companies from Spain, Italy, the UK, Chile, Belgium, Canada and the US to enter the Argentine gas transport and distribution market in association with Argentine companies. Residual government holdings in these companies are being sold off gradually.

## Licences, exclusive rights, open access and eligible consumers

Transmission and distribution companies were issued licences to operate the existing systems for a term of 35 years, renewable for an additional 10 years, after evaluation of the company's performance. At the end of this 35- or 45-year period, a competitive tender must be held for the licence. The incumbent will have the option of matching the best bid. The licences establish the principal rights and obligations of licensees, the terms and conditions of service, operating and safety standards, and penalties for cases of non-compliance. The licensing system is administered by Enargas, but ultimate authority rests with the federal government.

Transmission and distribution companies are granted exclusive rights for the commercial operation of the existing systems, but while distribution companies also have a monopoly for developing the network in their geographical areas, transmission companies do not.

---

13. Except for upstream gathering lines owned by producers.

Thus, a new market entrant may build a new high-pressure line anywhere in the country, but he may not build medium- and low-pressure links in areas covered by a licensed distributor. The split of GdE's transmission network into two companies was designed to promote competition between the two gas-transmission companies, by giving both of them access to the different sources of gas and to the main market of Greater Buenos Aires.

Transmission companies must grant non-discriminatory open access to all producers, retailers, distributors and eligible customers. To ensure non-discrimination among clients, transmission companies are not allowed to trade in gas. Gas producers, storage companies, retailers and consumers who contract for gas directly may not hold a controlling stake in transmission or distribution companies, and vice-versa. Distributors are not allowed to own a controlling stake in transmission companies, and vice-versa.

Distribution companies have exclusive rights for developing the network over their geographical areas, but monopoly rights are limited to clients consuming less than 5,000 cubic metres per day.[14] Those who consume more than this ("eligible consumers", mainly industrial users and power generators) can contract their gas in the wholesale market, i.e. they have the option of buying their gas directly from a producer or a trader, and can negotiate access to the transportation and distribution network, thus bypassing the distribution company ("commercial bypass"). Eligible consumers can also physically bypass the distribution network by building a direct link to the transmission network ("physical bypass"). In both cases, they need to negotiate with the transmission companies for firm or interruptible service.

However, most eligible consumers who buy their gas directly from producers opt to negotiate a bundled transmission-and-distribution service with the local distributor, because distributors generally offer rates and conditions which are considered more attractive. Direct gas transactions between producers/traders and eligible customers grew 40% between 1997 and 2000, reaching 32.6 mcm/d in 2000, equivalent to 38% of total gas demand.

## Prices and tariffs

The tariffs paid by the final consumer comprise three segments: the price of gas at the wellhead, the transport tariff and the distribution tariff. Wellhead gas prices were liberalised on 1 January 1994 (Decree 2731/93). The removal of price controls, which had previously equalised prices across the five producing basins, led to a divergence in prices reflecting the netback value of gas delivered to the main consuming area of Greater Buenos Aires. On average, effective prices (under prevailing contracts) rose by 15% between December 1993 and December 1995, then remained relatively stable until December 1999, rising again by 8-10% in 2000.

As the dominant supplier of natural gas in Argentina, Repsol-YPF effectively sets the price of gas at the wellhead and acts as a price leader for the market as a whole. However, the company does not seem to have used its market power to increase the wellhead price unduly, possibly for fear that, if it did so, the government might seek to correct

---

14. Initially, eligible customers were defined as those consuming more than 10,000 cubic metres per day. *Enargas* lowered the threshold to 5,000 cubic metres per day in 2000, thus extending access to the wholesale market.

the lack of competition in oil and gas production and supply. Despite the increase in prices in the two years following the removal of price controls, average wellhead prices and pre-tax consumer tariffs remain well below those in other major gas-consuming countries in North America and Europe.

The rules for determining the level and structure of distribution and transmission tariffs are set out in the Natural Gas Act and in the licences. Transportation and distribution tariffs are established for each company by Enargas on the basis of the cost of providing the service plus a reasonable rate of return on assets. The regulator effectively sets maximum tariffs ("price caps"), and transporters and distributors are free to offer lower tariffs to their clients. However, to ensure transparency and non-discriminatory treatment, transmission and distribution companies are required to publish their tariffs for different services and customer categories. The price caps are adjusted every six months for inflation.[15] Every five years, the tariffs are reviewed by Enargas. In this review, efficiency improvements and additional investment are factored in and fixed for five years, as a way for companies and consumers to share efficiency gains.

Distribution companies have two different prices. One is the bundled rate paid by final customers who are unable to choose their suppliers. This is made up of three elements: the price of gas at the point it enters the transportation system, transportation charges and distribution charges. Distributors are permitted to pass gas-purchase costs and transmission charges on to the consumer ("pass-through"). Occasionally, however, Enargas has blocked some distributors from passing on the full cost of buying gas, on the grounds that the distributors could have negotiated lower prices with producers in different basins. Cross-subsidies among customer categories are forbidden.

## Promoting competition

Promoting competition in gas supply and demand is one of the objectives of the 1992 Natural Gas Bill and has been a major endeavour for Enargas during its first ten years of existence.

The spot market, for trading short-term volumes of gas, has been slow to develop, but it has recently grown in importance. In 1995, Enargas established a "stick-and-carrot" incentive system (Decree 1020) to promote use of the spot market by distributors in order to reduce their gas-purchase costs (thus lowering gas prices for captive consumers). Under this system, Enargas establishes "reference prices". If the distributors are able to negotiate gas prices below the reference prices, they may retain half of the gain. If they buy gas at a higher price they can only pass half the difference on to the consumer. Most spot trading is done by distributors. In 2000, spot transactions covered on average 10% of all gas deliveries, with significant differences across basins and for summer and winter.[16] It can be observed that spot and contract prices are converging.

In 1997, Enargas issued a resolution (419/97) aimed at creating a secondary market in transmission capacity in order to promote more efficient use of the transmission network. Holders of firm transmission capacity can sell any unused reserved transmission

---

15  The adjustment for inflation was suspended in 2001, and eliminated in January 2002, following the devaluation of the peso.
16. In mature gas markets in North America and Britain, spot trade is significantly greater, exceeding the volume of gas actually delivered, due to reselling.

capacity. Prices for released capacity are to be established by market forces, but they are capped by the maximum tariffs for primary capacity, which are regulated by Enargas. This cap is designed to discourage distributors from deliberately reserving transmission capacity above their needs, so as to sell it on the secondary market. Due to lack of initial activity on this market (55 transactions between 1998 and 2000), in 2000 Enargas announced changes and new incentives, including the removal of the price cap, to encourage growth in this market.

Also in 1997, Enargas defined the role of the gas trader or retailer (Resolutions 421 and 478) who can buy and sell gas or transportation capacity, introducing some competition to distribution companies. In the period between April 1999 and March 2000, traders accounted for 13% of total gas supply.

**Gas exports and imports**

Imports of gas are authorised by law and do not require prior approval. Companies wishing to export gas must, however, obtain an export licence from the Secretariat of Energy. Until February 2001, Resolution N° 299 of the Secretariat of Energy set out the administrative procedure and conditions for issuing natural gas export licences. A distinction was made between long-term and short-term export authorisations. The resolution sought to balance the need to ensure adequate domestic supply with the benefits of international trade.

This resolution was replaced by a new one in February 2001, providing for automatic approval of gas-export requests, with the aim of promoting further exploration and development of the Argentine gas reserves. The government made this decision because the long-term gas supply outlook had improved. In particular:

■ gas reserves in Argentina and in Bolivia had increased;

■ agreements for gas integration between Bolivia and Argentina had been put in place;

■ supply from other areas with important reserves, such as Venezuela and Peru, appeared to be feasible in the medium-to-long term.

## THE CONSEQUENCES OF THE ECONOMIC AND FINANCIAL CRISIS ON THE GAS SECTOR

Argentina's economic and financial difficulties are having a major impact on the energy sector. The first companies affected by the *pesification* of the economy and the sharp devaluation of the peso were the gas and electricity distribution companies. A clause of the Economic Emergency Law, aimed at preventing inflationary pressures, invalidated the pegging of tariffs to the dollar, which was established in the concession contracts. Largely indebted in US dollars on the international markets, these companies saw their tariffs converted to pesos on a 1-to-1 basis. The government called for a renegotiation of the concession contracts. Meanwhile, the distribution companies, already debilitated by years of recession and affected by an increasing incidence of unpaid bills, are accumulating huge debts, which their international parent companies are unwilling to cover. The first to declare default on its debt payment was Metrogas, Argentina's largest natural gas distribution company, which supplies over 1.8 million customers in Greater Buenos Aires.

The situation is no less difficult in the upstream sector, particularly affecting those companies whose production mix leans on natural gas, where sales are mainly on the domestic market. With the devaluation of the peso, the wellhead price – which is determined by the spot price in Buenos Aires – has been falling drastically, in some cases below production costs. In addition, distributors have been falling behind on payments to producers, but producers are obliged by the regulator to continue supply. Contrary to expectations, exports to Chile do not contribute to alleviating the financial difficulties of gas producers, because the price of a large part of the gas exported to Chile is also indexed to the spot price in Buenos Aires.

The situation is only slightly better for oil producers, who have the opportunity to sell their product on the international market at international prices, if domestic prices are not attractive. Exports, however, are subject to a new export tax (20% on crude and 5% on products), imposed by the government as part of its emergency budgetary measures. In addition, in an attempt to keep domestic oil product prices from rising to international levels, the government is planning to implement restrictions on exports. This could lead to the re-emergence of an old problem which affected the Latin American energy industries in the 1980s: the conflict between policies aimed at controlling inflationary pressures and the liberalisation of energy prices as a precondition to viabilise new investments.

## CHALLENGES AND UNCERTAINTIES

At this point, it is very difficult to assess the full impact of the Argentine crisis on the country's gas sector. Domestic gas demand is stagnating and new investment in gas exploration, transportation, distribution and power generation are all on hold.

Gas exports to Chile still remain strong, as Chile's economy has been relatively unaffected by the Argentine crisis. Indeed, the *pesification* of the Argentine economy and the lower internal demand for gas in Argentina have lowered the price that Chileans pay for Argentine gas, thus contributing to increasing demand. As for exports to Brazil, there is much uncertainty over the volume of gas that will eventually flow from Argentina to Brazil. The evolution of Argentine gas exports to southern and southeastern Brazil will depend on the evolution of gas demand in the Brazilian power sector, on the growth of gas production in Brazil and on competition with Bolivian gas.

Over the longer term, there is also concern about Argentina's ability to maintain supply above internal demand and exports already committed to Chile. Here there seems to be a divergence of opinions between the government and the industry. The government is taking a conservative position, arguing that current reserves should not be committed to additional export projects, since even at the current level of demand the reserve-to-production ratio is only about 20 years. The industry, on the other hand, argues that there are good prospects to expand reserves and increase production because the Argentine producing basins are relatively new, and most sedimentary basins are still unexplored. There are, however, problems of location and cost, as much of the unexplored Argentine acreage is located in the remote southern part of the country and in difficult offshore areas. Major investments would be required to strengthen transmission between the south and the centre of Argentina. Until the country recovers from the current economic and financial crisis, new investments in exploration and production and in transmission are unlikely to materialise.

# CHAPTER 6
# BOLIVIA

With proven natural gas reserves that grew seven times since 1997, Bolivia is now the second holder of proven natural gas reserves in South America after Venezuela, but the first in terms of non-associated gas. Bolivia's geographical position and abundant gas resources make it ideally placed to become South America's gas hub and play a central role in Southern Cone energy integration. Thorough reforms have transformed the country's energy sector from a state-owned preserve to a predominantly private industry. The country's domestic energy market is small, but the prospect of supplying the region's largest and rapidly-growing energy market – Brazil – has attracted a large number of foreign companies. The issue facing Bolivia now is how to monetise its gas reserves. Exports to Argentina and Chile are being explored, as well as LNG exports to Mexico and the US.

## Bolivia at a glance, 2000

|  | Bolivia | Share in region |
| --- | --- | --- |
| Surface area: | 1,098,580 km² | 6.3% |
| Population: | 8,3 million | 2.4% |
| Capital: | La Paz | |
| Currency: | Boliviano ($B) | |
| GDP*: | US$19 billion | 0.8% |
| Total primary energy production: | 5.9 Mtoe | 1.0% |
| Primary energy consumption: | 4.9 Mtoe | 1.2% |
| Primary gas consumption: | 1.3 Mtoe | 1.5% |
|  | **Bolivia** | **Regional average** |
| Per capita GDP*: | US$2,286 | US$6,821 |
| Per capita primary energy cons.: | 0.59 toe | 1.15 toe |
| Per capita electricity consumption: | 391 kWh | 1,708 kWh |

\* *Gross domestic product (GDP) expressed in 1995 prices and PPP.*

## BACKGROUND

### Economic development

Bolivia is the fifth-largest South American country in terms of surface area and one of only two land-locked nations in the continent. It shares borders with Brazil to the north and northeast, Paraguay to the southeast, Argentina and Chile to the south and Peru to the west. Just over eight million people live in an area of approximately 1.1 million square kilometres. Population density is the lowest in South America, at 7.1 people per square kilometre. Urbanisation is also lower than the regional average: 67% of the population live in urban centres, compared to 79% for South America as a whole. However, in the last ten years, urban population has been growing considerably faster

(4.3% per year) than total population (2.4% per year). According to the UN Economic Commission for Latin America and the Caribbean (UN-ECLAC), total population growth is expected to continue at an average of 1.9% p.a. for the next 20 years.

Bolivia is one of the poorest countries in South America. In 2000, per capita GDP stood at US$2,286 (expressed in 1995 prices and adjusted for Purchasing Power Parities - PPP). This is about one-third of the regional average of US$6,800. The country is going through a period of economic difficulties. Economic growth was 2.4% in 2000 and 1.2% in 2001,[1] better than the 0.4% of 1999, but still far below the 4.5% average from 1990 to 1998. The 1999 slowdown was caused primarily by a contraction of almost 15% in export earnings for the year. This was due in large part to the termination of natural gas exports to Argentina. The gradual increase of gas exports to Brazil will contribute to renewed economic growth in coming years. Official projections for GDP growth in 2002 vary between 2.5% and 3.5%; ECLAC, however, forecasts a 1.8% growth.

## Economic reforms and privatisation

For most of its history, Bolivia has suffered from political and economic instability. In the decade after civilian rule was restored in 1982, it managed to establish democracy and defeat hyperinflation. Bolivia was the second country in South America (after Chile) to launch a series of IMF-backed economic reforms in the mid-1980s. Reforms included the elimination of import and investment permit requirements and the opening to private participation of economic activities previously reserved to the state.

A second wave of market-oriented reforms was launched under the first government of President Gonzalo Sánchez de Lozada (1993-1997).[2] These included the privatisation of most public enterprises and the "capitalisation" of five major state-owned companies in the energy, transport and telecommunication sectors.

"Capitalisation" is the name of a unique privatisation scheme devised by Bolivia. Under this scheme, the state ceded 50% of its shares and management control of state companies to foreign investors, in return for explicit investment commitments. These commitments totalled US$1.7 billion to be spent within seven years. The remaining 50% of the shares were transferred to a Bolivian pension fund.[3]

The capitalisation scheme differs from other privatisation methods in several key aspects:

■ The government does not sell off the state-owned company, but sets up "mixed capital corporations" (MCCs), to which a private partner contributes a 50% capital investment;

■ The private partner's contribution increases the value of the MCC substantially;

■ The foreign investment is not used to solve budget deficit problems, but is used instead to expand the production apparatus or the capital stock of the companies.

The capitalisation programme attracted unprecedented levels of foreign investment to Bolivia and was hailed as a model for privatisation worldwide. However, despite the

---

1. ECLAC (2002), "Economic Survey of Latin America and the Caribbean 2001-2002".
2. Mr. Sánchez de Lozada was re-elected President of Bolivia in August 2002.
3. All Bolivians over the age of 21 as of 31 December 1995 (approximately 3.4 million people or 50% of the total population) are eligible to receive benefits from the government's share of the capitalised companies.

international praise, the scheme was strongly criticised in Bolivia, where it was the frequent target of street protests. Critics argue that the capitalisation of state-owned companies did not promoted economic activity, nor did it generated much direct or indirect employment, as the capitalised companies undertake capital-intensive activities requiring raw material, equipment and expertise which are not produced or available in Bolivia.

## Foreign investment

Foreign direct investment (FDI) in Bolivia reached an historic high in 1999, totalling nearly US$1,016 million, equal to 10.5% of GDP. Of total FDI, 67% went to the hydrocarbons sector, as a result of the privatisation process. FDI fell in 2000 and 2001, but remained strong, with increasing activity in the oil, gas and electricity sectors as the driving force.

To promote foreign investment in the energy sector, the Bolivian government introduced the *Ley Corazón* ("Heart Law"), which removes restrictions on foreign ownership of property within 50 kilometres of the borders. This law, in combination with another one that creates tax-exempt areas for energy-export projects, is intended to encourage companies to build gas-fired power plants for export markets.

There are no exchange controls, and no restrictions on the repatriation of capital, dividends, interests, or any other remittances abroad.

## Regional integration

Bolivia is a member of the Andean Community (*Comunidad Andina de Naciones* – CAN), along with Colombia, Ecuador, Peru and Venezuela. These five countries established a free-trade zone in 1993 and are working towards the establishment of a common market by 2005. Since 1996, Bolivia is also an associate member of the Southern Cone Common Market (*Mercado Común del Sur* – Mercosur), the commercial trading block comprising Argentina, Brazil, Paraguay and Uruguay, with Chile as another associate member. In 1998, the five CAN countries and the four Mercosur members started negotiations towards the creation of a free-trade area between the two blocks.

Bolivia's relations with Chile have been tense since the end of the 19th century. In 1884, Bolivia lost the "Pacific War" with Chile and, with it, its entire coastal area. The conflict erupted after Bolivian President Hilarion Daza rescinded a contract with a Chilean company to mine nitrate deposits in Atacama, which was once a province of Bolivia. Chile took the port of Antofagasta in reprisal and war was formally declared in 1879. Chile also declared war on Peru, which had a defensive alliance with Bolivia. Chile not only defeated its opponents, but took control of the entire Bolivian coast, the Antofagasta region with its nitrate, copper and other mineral industries. A treaty in 1904 made this situation permanent, leaving Bolivia a landlocked nation. Due to Chile's refusal of Bolivian demands to regain an outlet to the Pacific, diplomatic relations between the two countries have been suspended since 1962, with just a brief resumption between 1975 and 1978. Since that date, relations have been conducted at the consular level and through various joint committees.

In 2001, the two governments started informal discussions about a natural gas export project involving the construction of a gas pipeline from Bolivia to a Pacific port (in Chile or Peru) and the construction in that port of a liquefied natural gas (LNG) plant. This has reawakened Bolivia's aspirations to regain access to the sea and raised hopes that a deal could be reached with Chile over the sovereign control of part of the Chile's northern coast. But so far the Chilean government has been firm about keeping the two issues well separated. The debate became politicised in Bolivia in the months preceding the 2002 presidential election. Protesters took the streets in La Paz in July 2002 to express opposition to an LNG deal with Chile.

## OVERVIEW OF THE BOLIVIAN ENERGY SECTOR

Bolivia is well endowed with hydrocarbon resources. Its proven oil and gas resources increased substantially in the last five years, following the opening of the sector to private investors in the mid-1990s and as a result of aggressive exploration aimed at supplying export markets. The country has long been a gas exporter, first to Argentina and now to Brazil. It is also a small exporter of crude oil to Chile and is now gearing up to export electricity. At present there are no interconnections between the Bolivian electricity system and those of neighbouring countries, but the Bolivian government is working on electricity-export agreements with Peru, Chile and Brazil.

**Figure 6.1** Primary energy supply in Bolivia by fuel, 1975-2000

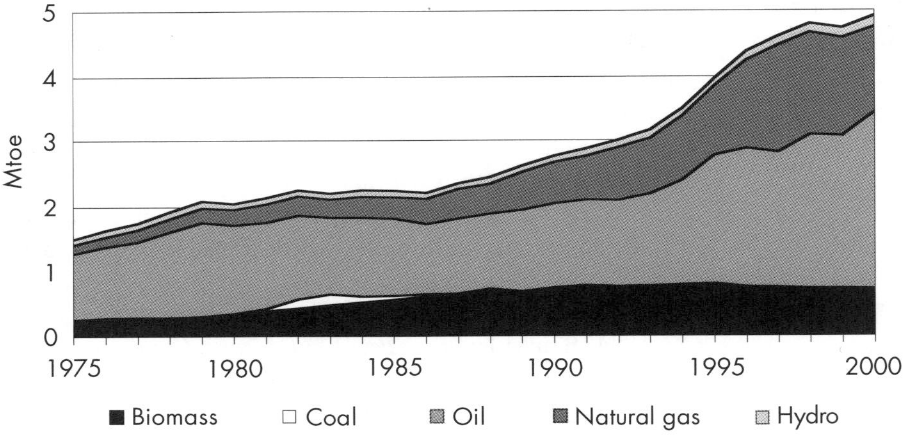

Source: IEA.

Bolivia's domestic energy market is relatively small compared to those of its Southern Cone neighbours, Argentina and Brazil. In 2000, Bolivia's total consumption of primary energy (TPES) stood at 4.9 Mtoe, just 1.2% of total South American TPES. After growing 6.8% per year between 1990 and 1998, Bolivia's primary energy consumption decreased by 2.6% in 1999 because of the economic slowdown, but grew again by 4.1% in 2000.

Energy consumption is low, partly because of the country's small population, and partly because of Bolivia's low per capita energy and electricity consumption: at 0.59 toe/person and 391 kWh/person, respectively, these indicators are roughly one-half and less than one-fourth of the corresponding averages for South America. According to official data, 74% of rural households and 22% of urban households had no access to electricity in 2001. In 1998, the government launched an ambitious plan to electrify rural areas.

The National Rural Electrification Plan (*Proner*) involves the expansion of grid connections and the development of alternative energy sources for isolated communities, such as solar and biomass-fuelled power generators. As a result, rural electrification grew from less than 15% of households in 1997 to 26% in 2001.

Electricity production grew by 6.4% per year between 1990 and 2000. Approximately one-third of the country's generating capacity is hydro and two-thirds is thermal, mostly gas-fired. In 2000, total electricity generation stood just under 4,000 GWh, of which about half was produced from natural gas and half from hydropower. The share of gas-fired generation has grown rapidly until the mid-1990s and has now stabilised around 45-50% (*Figure 6.2*). Bolivia has substantial untapped hydro resources.

**Figure 6.2** Electricity generation in Bolivia, 1975-2000

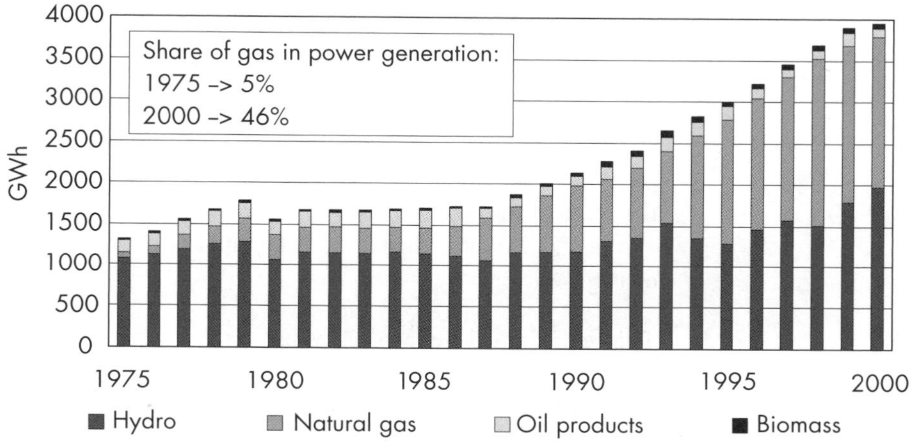

Share of gas in power generation:
1975 -> 5%
2000 -> 46%

GWh

1975    1980    1985    1990    1995    2000

■ Hydro    ■ Natural gas    □ Oil products    ■ Biomass

Source: IEA.

While Bolivia's domestic energy market is expected to expand in the next decade, driven by economic growth and industrialisation, exports to neighbouring countries will remain the main driver for the development of the Bolivian energy industry.

**Energy-sector structure and institutions**

Reform in the Bolivian energy sector began with the 1994 Capitalisation Law, which paved the way for the partial privatisation of five state-owned companies, among which were the electricity company ENDE and the oil and gas company *Yacimientos Petrolíferos Fiscales Bolivianos* (YPFB). Bolivia's energy sector is now characterised by a number of mixed-capital companies controlled by private shareholders.

The Ministry of Hydrocarbons and Energy is responsible for formulating and implementing the country's energy policy. It promotes private investments and exports, designates areas for bidding and establishes wellhead prices.

The body responsible for the regulation of the oil and gas sectors is the *Superintendencia de Hidrocarburos* (SH). The SH is one of five independent government agencies that were created to oversee the sectors privatised since 1994. These five *superintendencies* – for hydrocarbons, electricity, water, telecommunications and transport – are part of the Sectorial Regulatory System (SIRESE). These superintendencies are autonomous bodies financed by taxes imposed on companies operating in their respective sectors. The

Superintendent General of SIRESE acts as a "regulators' regulator", controlling and supervising the five superintendencies and acting as the first instance of appeal against any decision made by an individual superintendent.

## NATURAL GAS RESERVES AND PRODUCTION

Bolivia has been producing gas since the 1960s, but production increased sharply in the early 1970s, when Bolivia started exporting gas to Argentina. Bolivian exports to Argentina started in 1972 and precipitated a "gas rush" in Bolivia, with very successful exploration results. Unfortunately for Bolivia, huge reserves were subsequently found in Argentina, and Argentine demand for Bolivian gas fell well short of Bolivia's supply capacity, leading to more than two decades of quotas on Bolivian production.

The opening of the country's energy sector to private investment in 1994 and the perspective of supplying gas to Brazil through the Bolivia-Brazil pipeline (*Gasbol*) have attracted many large regional and international companies to Bolivia's E&P sector. Aggressive exploration efforts have led to the discovery of several large gas fields in the south of the country. As of 1 January 2002, Bolivia's proven gas reserves stood at 775 bcm, 15% more than in 2001 and more than seven times the January 1997 level of 106 bcm.[4]

This figure is likely to increase further in the next few years, following recent discoveries not yet proven. At the beginning of 2002, estimated probable and possible reserves stood at 706 bcm and 704 bcm respectively. While there is significant potential for new discoveries, further exploration is unlikely to be undertaken in the near term, as discoveries have outpaced demand. The recent evolution of Bolivia's proven, possible and probable reserves is shown in *Figure 6.3*.

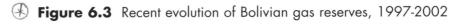 **Figure 6.3** Recent evolution of Bolivian gas reserves, 1997-2002

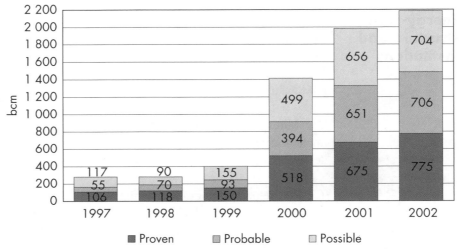

*Source: YPFB.*

---

4.   Official data from YPFB. Other estimates of Bolivia's proven reserves are even higher.

This remarkable increase in gas reserves over the past five years is attributable largely to drilling discoveries by TotalFinaElf on its Tarija Oeste block, by Repsol-YPF on its Caipipindi block and by Petrobras on its San Antonio and San Alberto blocks. Bolivia has now four separate fields each with over 5 tcf (142 bcm) of reserves, which are defined as "giant" fields by international standards: San Alberto, San Antonio, Itau and Margarita. The San Alberto field began production in 2001, while the other 3 fields are still under development. All are located in the Tarija district, in the south of the country, close to the border with Argentina. As a result, the district of Tarija now accounts for nearly 90% of the country's proven and probable reserves (see *Table 6.1*). The location of Bolivian gas reserves is shown in *Map 9* at the end of the book.

**Table 6.1** Bolivian natural gas reserves by area, as of 1 January 2002

|  | Proven | Probable | Total | |
|---|---|---|---|---|
|  | bcm | bcm | bcm | % |
| Cochabamba | 79 | 59 | 138 | 9% |
| Chuquisaca | 19 | 9 | 29 | 2% |
| Santa Cruz | 16 | 3 | 19 | 1% |
| Tarija | 661 | 635 | 1 295 | 87% |
| **Total** | **775** | **706** | **1 481** | **100%** |

*Source: YPFB.*

Since 2001, Bolivia has replaced Argentina as the second most important holder of natural gas reserves in South America after Venezuela. Significantly, Bolivia has the largest non-associated gas reserves in the continent. Bolivian authorities, who initially feared that the country did not have sufficient reserves to fulfil its export contracts with Brazil, are now concerned that Bolivia's rising production capacity will rapidly outstrip Brazil's demand, and they are already looking for new export markets (Argentina, Chile, LNG to Mexico and the US).

According to official sources, gross production reached 7.15 bcm (19.6 mcm/d) in 2001, a 25% increase over 2000 production. Of these 7.15 bcm, 24% was reinjected,

**Figure 6.4** Natural gas gross and marketed production in Bolivia, 1975-2000

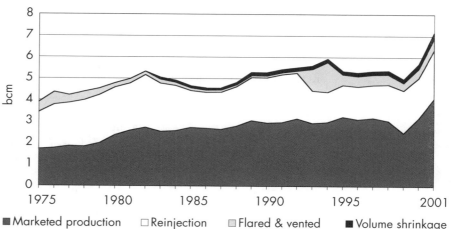

*Source: Cedigaz.*

3% was flared or vented, 2% was converted to liquids,[5] 3% was used as fuel in the E&P sector and the remaining 68% was commercialised. *Figure 6.4* shows the trend of Bolivia's gross and marketed[6] natural gas production.

## Main players

*Empresa Petrolera Andina S.A.* (Andina) was formed in 1997 as part of the restructuring and capitalisation process of YPFB's upstream assets. A consortium of the Argentine companies, including YPF (now Repsol-YPF), *Perez Companc* and *Pluspetrol*, acquired 50% of the company in 1997. Company employees and private Bolivian pension funds control the remaining 50%. Repsol-YPF subsequently bought Perez Companc's 20.25% stake[7] and Pluspetrol's 9.5% stake in Andina. Andina has a 50% stake in the San Alberto and San Antonio blocks, operated by Petrobras, which together account for 583 bcm. Altogether, Andina owns about a quarter of Bolivia's total proven and probable natural gas reserves.

*Empresa Petrolera Chaco S.A.* (Chaco) is another company that was set up in 1997 as part of the capitalisation of YPFB's upstream assets. UK's BP and Argentina's *Bridas Corp.* hold 50% of the shares and have administrative control of the company, Bolivian pension funds own 47%, and employees of YPFB own 3%. Chaco owns 4% of Bolivia's proven and probable reserves.

Britain's *BG* and France's *TotalFinalElf* (through its wholly-owned affiliate of *Total Exploration Production Bolivie* - TEPB) each own about 15% of Bolivia's proven and probable gas reserves, with 238 bcm and 208 bcm respectively. TEPB began exploration activities in Bolivia in 1997. TEPB is the operator of the block "XX Tarija Oeste", where it discovered the giant Itau field. TEPB has a 41% stake in the Itau field; the remaining stakes are owned by ExxonMobil (34%) and BG (25%). TEPB is also associated with Andina and Petrobras in the San Alberto and San Antonio blocks. BG has stakes in the Itau field and in the Caipirendi block.

Brazil's state oil company *Petrobras* participates in eight of Bolivia's 60 exploration blocks and is the main operator in four of them. It is also the operator of the giant San Alberto and San Antonio fields. Petrobras has shares in both the Bolivian and Brazilian portions of the Bolivia-Brazil natural gas pipeline, and has priority access to the pipeline (see *Chapter* 7).

Among the foreign companies that have bought assets in Bolivia, *Repsol YPF*, through its affiliates Andina and Maxus, is the largest company in Bolivia in terms of oil and gas reserves, production and sales. According to recent estimates, on 1 January 2002, Repsol-YPF's proven and probable gas reserves in Bolivia amounted to 325 bcm.

Other companies with smaller interests in the Bolivian gas market are US-based *ExxonMobil, Pan American Energy* (a subsidiary of BP, 50%, and Bridas Corp., 50%), US-based *Vintage Petroleum,* and Argentina's *Pecom Energía* (a wholly-owned subsidiary of Argentina's Perez Companc, of which Petrobras recently bought a 58.63% stake).

---

5. Official data from YPFB. Cedigaz gives a different distribution: 32% reinjected, 8% flared or vented, and 4% of volume losses, due to purification and extraction of natural gas liquids (NGL).
6. See Glossary for definitions of gross and marketed gas production used by IEA (and Cedigaz).
7. In February 2001, as part of a US$434.5-million asset-swap agreement.

*Table 6.2* shows the natural gas reserves held by each company as of 1 January 2002, while *Table 6.3* gives an overview of the gas production of each operator in 2001.

**Table 6.2** Bolivian natural gas reserves by company, as of 1 January 2002

| Company | Blocks | Participation in blocks | Proven and probable reserves | |
|---|---|---|---|---|
| | | | bcm | % |
| Andina S.A. | Ex-YPFB blocks | 100% | 76 | |
| | San Alberto | 50% | 169 | |
| | San Antonio | 50% | 122 | |
| | | | 367 | 24.8% |
| BG | "XX Tarija Oeste" | 25% | 74 | |
| | Caipipendi | 38% | 144 | |
| | Other blocks | 100% | 20 | |
| | | | 238 | 16.1% |
| TotalFinaElf | "XX Tarija Oeste" | 41% | 121 | |
| | San Alberto | 15% | 51 | |
| | San Antonio | 15% | 37 | |
| | | | 208 | 14.0% |
| Petrobras | San Alberto | 35% | 118 | |
| | San Antonio | 35% | 86 | |
| | | | 204 | 13.8% |
| Maxus / Repsol-YPF | Caipipendi | 38% | 144 | 9.7% |
| ExxonMobil | "XX Tarija Oeste" | 34% | 100 | 6.8% |
| PanAmerican Energy | Caipipendi | 25% | 96 | 6.5% |
| Chaco S.A. | Ex-YPFB blocks | 100% | 59 | 4.0% |
| Vintage Petroleum | | 100% | 31 | 2.1% |
| Pecom Energía S.A. | Caranda - Colpa | 100% | 22 | 1.5% |
| Other companies | | | 11 | 0.7% |
| Total Bolivia | | | 1,481 | 100% |

*Source: YPFB.*

**Table 6.3** Bolivian natural gas production by operator, 2001

| | Gross production | Marketed production | |
|---|---|---|---|
| | mcm | mcm | % |
| Andina | 2,449 | 1,301 | 27% |
| Chaco | 1,787 | 1,115 | 23% |
| Petrobras | 1,041 | 997 | 20% |
| Maxus | 561 | 250 | 5% |
| BG | 447 | 432 | 9% |
| Pecom Energía | 431 | 380 | 8% |
| Vintage Petroleum | 381 | 367 | 7% |
| Pluspetrol | 51 | 50 | 1% |
| Matpetrol | 4 | - | 0% |
| Canadian | 3 | - | 0% |
| Dong Wong | 0 | - | 0% |
| **Total** | **7,154** | **4,891** | **100%** |

*Source: YPFB.*

## NATURAL GAS DEMAND

Primary natural gas demand stood at 1.7 bcm in 2001 (equivalent to 1.4 Mtoe on a net calorific basis), accounting for 38% of the total marketed gas production. Gas consumption has increased 11% annually since 1971. In 2000, 38% of natural gas supply was used by the power-generation sector, 13% by the oil and gas upstream sector, 19% is the energy sector's own use and distribution losses, 26% by the industrial sector, and just 3% by other end-use sectors (transport, residential, commercial).

The share of gas in total primary energy rose from 6% in 1971 to 36% in 1997, though it has dropped since then, to 27% in 2000. In 2000, gas accounted for 46% of the power-generation output, 38% of the energy used by industry, 9% of the energy used by the commercial and service sectors, 2% of the energy use in transport and less than 1% of residential energy use. At the end of 2000, there were 11,000 household and industrial users, most of them in Cochabamba.

**Figure 6.5** Natural gas demand in Bolivia, 1971-2000

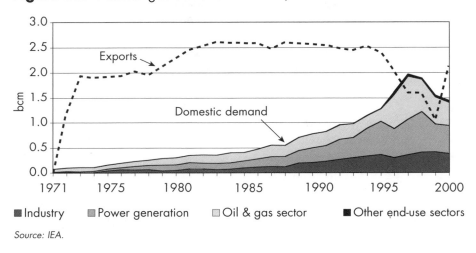

Source: IEA.

Increased use of gas in the power generation is hindered by the abundance of hydropower, though there are plans to build gas-fired power plants near the border with Brazil to supply the Brazilian grid. There is still scope for increased gas penetration in all end-use sectors. As in other South American countries, the replacement of other fuels by natural gas is a long-term goal for the Bolivian government. Plans to phase out subsidies for domestic LPG could give consumers an incentive to switch to natural gas. The planned privatisation of the natural gas distribution networks – still controlled by the government-owned company YPFB – may help bring about the necessary investment to expand the network and connect new consumers.

Doubts remain, however, as to whether the small domestic market can attract private investors. All in all, the primary destination for Bolivian gas will remain the export market, either in the form of piped gas and gas-generated electricity to neighbouring countries, or in the form of LNG.

**Gas transmission and distribution**

Bolivia's gas transportation network covers the southern two-thirds of the country and consists of two interconnected systems (see *Map 9*). The north-western system connects the cities of La Paz, Oruro, Cochabamba and Santa Cruz via 280 kilometres of 12-inch, 16-inch and 24-inch pipelines from Rio Grande to the north. The southern system connects Yacuiba, Sucre and Potosi and extends northward to Rio Grande, where the Bolivia-to-Brazil (Gasbol) pipeline starts. The southern gas system consists of 442 kilometres of 24-inch pipelines. The domestic markets for gas are located mainly in the cities listed above.

The country's domestic gas transportation system is owned and operated by *Transredes S.A.*, a company jointly owned by Enron (25%), Shell (25%), Bolivian pension funds (34%), and Bolivian workers at Transredes and YPFB (16%). Transredes also has a majority participation and is the operator of two other Bolivian gas transportation companies: GasTransboliviano S.A and Gas Oriented Boliviano Ltda.

*GasTransboliviano S.A.* owns and operates the Bolivian portion of the Bolivia-Brazil pipeline. Transredes has a 51% participation in the company. The other investors include: Petrobras (9%), Enron (4.5%), Shell (4.5%), BG Group (2%), TotalFinaElf (2%), El Paso (2%) and Bolivian interests (25%).

*Gas Oriented Boliviano Ltda.*, which owns and operate the San Miguel-Cuiabá pipeline, is a joint venture of Enron and Shell.

There are currently five distribution companies in Bolivia, serving around 11,000 domestic and industrial consumers. Four of the companies are operated by private interests: Emcogas in Cochabamba (the largest), Emtagas in Tarija, Emdigas in Sucre and Sergas in Santa Cruz. In addition, YPFB still operates the distribution network in La Paz, Ouro and Potosí. The government has recently unveiled a plan to add 250,000 connections in the next five years.

## NATURAL GAS EXPORTS

Bolivia started exporting gas to Argentina in May 1972, under a 20-year contract. Argentina subsequently found abundant reserves of its own and exports from Bolivia to Argentina never exceeded 6 mcm/d. Nevertheless, the gas exports provided the Bolivian government with significant income, reaching US$380 million per year in the mid-1980s. The long-term contract with Argentina expired in 1992, but the two governments agreed to several consecutive extensions (although at a much lower price) to provide income to Bolivia until the start of exports to Brazil. The contract expired on 31 July 1999. All in all, Bolivia exported 53 bcm of natural gas to Argentina, worth about US$4.5 billion.

Currently, Argentine producer Pluspetrol exports small quantities of gas to Argentina from its Bolivian fields of Bermejo y Madrejones, situated close to the Argentine border. Large-scale exports to Argentina are unlikely to resume in the short to medium term, but are not excluded in the longer term, given Argentina's low reserve horizon.

Export to Brazil started in July 1999, under a 20-year contract between YPFB and Petrobras. Exports are expected to increase gradually to 30 mcm/d in 2003. So far, exports have lagged behind the agreed schedule because of problems in developing the gas market in Brazil.

In 2001, Bolivia exports 3,679 mcm of gas to Brazil and 43 mcm to Argentina. *Figure 6.6* gives an overview of the evolution of Bolivian gas exports.

**Figure 6.6** Bolivia's natural gas exports, 1991-2001

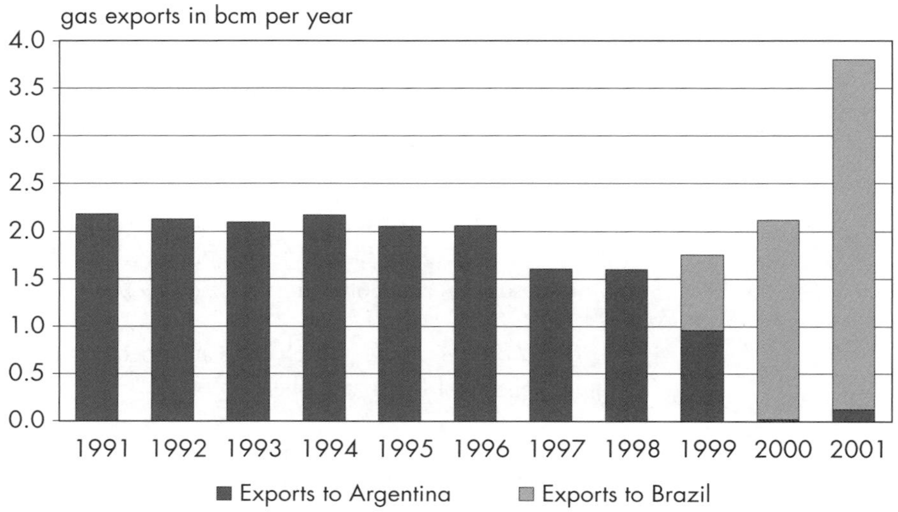

Source: YPFB.

**Cross-border gas pipelines**

**Bolivia-Argentina**

Until 1999, Bolivia's gas network was interconnected only with Argentina's, through the *Yabog* (Yacimientos-Bolivian Gulf) pipeline, a 541-km, 24-inch, 6-mcm/d trunk line running southward from Rio Grande to Yacuiba and Campo Durán. Two small pipelines owned by Argentina's Pluspetrol link Bermejo (Tarija, Bolivia) to Ramos (Salta, Argentina), and Madrejones (Tarija, Bolivia) to Campo Durán (Salta, Argentina).

**Bolivia-Brazil**

The first part of the Bolivia-to-Brazil (*Gasbol*) pipeline, extending 1,800 kilometres from Rio Grande, Bolivia, to São Paulo, Brazil, was completed in December 1998, and Bolivian gas started flowing to Brazil in July 1999.[8] When all compression stations are installed, the pipeline will have a throughput capacity of 30 mcm/d. The second part of the pipeline, a 1,100-km extension from São Paulo to Porto Alegre, Brazil, was completed in April 2000. With a total of 3,150 kilometres, the Gasbol is the longest pipeline in Latin America. Its total cost was estimated at US$2.5 billion.

According to the initial contract signed by Petrobras and Bolivia's YPFB, the pipeline is expected to reach its full capacity of 30 mcm/d by 2004 by adding additional compression. Exports gradually increased to around 7 mcm/d at the end of 2000, and reached 12.5 mcm/d in September 2002.

---

8. See also *Chapter 3* and *Chapter 6*.

A second line, known as Lateral Cuiabá was opened in 2000. This 620-km spur off the Gasbol pipeline runs from San Miguel in Bolivia to Cuiabá in Brazil, where it fuels a 480-MW thermal power station. This integrated project was developed by Enron and Royal Dutch Shell at an estimated cost of US$220 million. The pipeline is operated by Gas Oriented Boliviano Ltda.

Based on current trends, exports to Brazil and domestic consumption are forecast to require 300 bcm in dedicated reserves over the next 20 years. With recent discoveries, proven reserves now clearly surpass what can be used domestically or put through the Gasbol pipeline. Loops will need to be built on the Gasbol if volumes to Brazil are to be increased beyond the current maximum capacity of 30 mcm/d. Total and Petrobras are reportedly studying the feasibility of constructing a second gas pipeline to Brazil or doubling the capacity of the Gasbol pipeline.

**Other pipeline projects**

Overall consumption of gas in Brazil is expected to increase dramatically in the next decade. Petrobras announced in 2001 its intention to buy 10 mcm/d of Bolivian gas in addition to the already-contracted 30 mcm/d by 2004. This will require an expansion of the Gasbol pipeline's capacity.

In addition to increased exports to Brazil, Bolivia is considering exporting gas to Argentina, Paraguay and Chile. There were plans to build a pipeline to Paraguay, the *Trans-Chaco* pipeline, running from Vuelta Grande in Bolivia to Asunción in Paraguay. The pipeline would then be extended to southern Brazil, as the potential Paraguayan market is not big enough on its own to warrant the required investment. More recently, another project to link Bolivia, Argentina, Paraguay and Brazil was unveiled by the Brazilian government. The so-called Gas Integration (*Gasin*) pipeline would start near the gas fields in southern Bolivia and run through northern Argentina, where a spur would cross the border to Asunción, the capital of Paraguay. The pipeline would then enter Brazil at the western border of Santa Catarina state. According to initial plans, the feasibility study is to be concluded in 2002, and the pipeline could be ready to start operation in 2005.

**Prospects for LNG exports**

Due to Argentina's significant reserves and mature market, Chile's easy access to Argentina's main gas fields and the small size of Paraguay's potential market, opportunities for additional Bolivian gas exports remain limited. It is clear that Bolivian gas reserves greatly exceed current demand in the Southern Cone, and Bolivian producers have started to look for other ways to cash in on their reserves.

Current discussions centre on an LNG project to bring Bolivian gas to North American markets, proposed by British BG and its partners in the Margarita field, Repsol YPF and BP. The project, named Pacific LNG, would involve the transportation of gas across the Andes to a liquefaction plant on the coast of Chile or Peru, for onward shipping to Mexico and the United States. TotalFinaElf is studying a similar project but it is likely to join Pacific LNG if it goes ahead.

The Bolivian government has been holding discussions with Chile and Peru to secure access to the sea for its natural gas. The ports being considered are shown in *Map 9* at the end of the book. The Pacific LNG consortium favours a port in Chile, as the Chilean coastline is closer and the cost of the pipeline would be lower. However, the Bolivian government faces a very sensitive political choice, as anti-Chilean sentiment is still strong in Bolivia, due to the loss of its coastline to Chile after war in the 19th century. As of 1 December 2002, no decision had been made.

## GAS-SECTOR REFORM AND PRIVATISATION

Reform in the Bolivian gas sector began with the 1994 Capitalisation Law, which paved the way for the partial privatisation of five state-owned industries, among them Yacimientos Petrolíferos Fiscales Bolivianos (YPFB), the Bolivian state oil and gas company.

The first step of the reform was the reorganisation and unbundling of YPFB into several independent business units: two upstream units, a transport unit, two refining companies and a marketing company. The capitalisation of the upstream and transportation units took place in 1996 and 1997. Even before the privatisation of YPFB, about 25% of the country's natural gas production originated from fields operated by private companies. Today, all natural gas production is in the hands of private companies, which have shared-risk contracts with YFPB.

The government has also begun to privatise the country's natural gas distribution grids and is encouraging investors to build additional systems, but the process has yet to be completed. There are currently four private distribution companies in addition to the distribution assets still owned by YPFB. The divestiture of YPFB's remaining gas distribution lines has been announced and then delayed several times. It now seems that the network will be offered as a single concession, instead of being split into two or three parts, as initially planned.

Once YPFB has sold off all its residual oil and gas assets, the company's main role will be to collect the upstream rent (royalties) and to be responsible for international negotiations and the administration of contracts signed with foreign oil companies. In particular, YPFB is the administrator of the gas export contract with Brazil.

**The new legal and regulatory framework**

The new Hydrocarbons Law (No.1689), enacted in April 1996, defined the modalities for developing the natural gas industry, the characteristics of the regulatory agency (the *Superintendencia de Hidrocarburos*-SH) and the regulatory mechanisms governing exports and the domestic market. The law is designed to promote gas exports.

The regulator's main tasks include implementing the 1996 Hydrocarbons Law, setting gas transportation tariffs, granting concessions and licences, and ensuring that privatised companies do not engage in anti-competitive or monopolistic practices and that they abide by their contractual obligations.

The law established a clear separation between transmission, distribution and power generation, in order to ensure non-discriminatory access to the transmission network.

Transmission companies are not allowed to trade in gas, and they may not participate directly in distribution activities or in power generation.

Gas producers are allowed to build and operate transmission networks to transport their own and third parties' products, but they must keep separate accounts for their production and transmission activities. Any company may request a concession to build, maintain and exploit a transmission gas network. The concession, issued by the regulator, does not grant exclusive rights and applies only to the specific project approved. Transmission tariffs are established every four years. Transmission companies must grant third party access to their networks, when they have available capacity.

A distribution concession is required to distribute gas in a particular area. Any company, except transmission companies, may request a distribution concession. Companies must participate in a bidding process to obtain distribution concessions, which are issued by the regulator.

## CHALLENGES AND UNCERTAINTIES

Bolivia is strategically located to supply Brazil's southern, southeastern and centre-western states, where energy demand is concentrated. But gas demand has been slow to pick up in Brazil, mainly due to considerable delays in the construction of planned thermoelectric power stations.

In its endeavour to increase its exports to Brazil, Bolivia will certainly face tough competition from Argentina, which is also seeking to supply Brazil's growing market. But Argentina has a mature gas market with considerable gas demand, and the lack of investments in exploration and production means that the Argentine R/P ratio has fallen to under 20 years. Some analysts believe Argentina may become an importer of Bolivian gas in the longterm.

Whether Bolivia will become an LNG exporter will depend on many factors, including the economics of an LNG project which includes a long cross-border pipeline, the competition from potential Peruvian LNG exports and the demand for natural gas on the west coast of North America. Most observers agree that the United States' traditional sources of gas – domestic production and imports from Canada – will not be able to keep up with demand, which is expected to increase substantially over the next decade, especially for power generation. Some South American LNG is likely to be imported also by Mexico, either for re-export to the US or for supplying its rapidly-growing domestic market, where demand is quickly outstripping production capacity. Several projects are under way in Mexico and in the US to build new LNG-receiving terminals or reactivate terminals which had been mothballed. However, Bolivian LNG will have to compete on price with several other sources of gas.

# CHAPTER 7
# BRAZIL

Brazil is the largest energy market in South America, accounting for 40% of the continent's energy consumption. Historically, natural gas has contributed very little to Brazil's energy mix, although the country has significant hydrocarbon resources. A severe drought-induced power crisis in 2001 exposed the vulnerability of a system that relies on hydropower for 90% of its electricity supply. The government aims to increase the share of natural gas in the energy mix to 12% in 2010, up from the current 4%, mostly as a result of new gas-fired generation. The recent opening of the upstream oil and gas sector to foreign investment should boost domestic natural gas production, but most of Brazil's natural gas supply is expected to come from imports. Since 1999, imported gas from Bolivia has been flowing through the 3,150-km Bolivia-Brazil pipeline, and several other pipelines are under construction or planned to bring gas from Argentina. However, the development of the Brazilian natural gas market has been slower than expected. In particular, uncertainties surrounding the structure and regulation of the electricity market, as well as the inherent complexities of introducing gas-fired power generation in a hydro-dominated power-generation sector, have contributed to delaying several gigawatts of gas-fired power projects.

## Brazil at a glance, 2000

|  | Brazil | Share in region |
|---|---|---|
| Surface area: | 8,547,400 km² | 49% |
| Population: | 170 million | 49% |
| Capital: | Brasilia | |
| Currency: | Real/reais (R$) | |
| GDP*: | US$1,184 billion | 50% |
| Total primary energy production: | 142 Mtoe | 24% |
| Primary energy consumption: | 183 Mtoe | 46% |
| Primary gas consumption: | 7.5 Mtoe | 9% |
|  | **Brazil** | **Regional average** |
| Per capita GDP*: | US$6,949 | US$6,821 |
| Per capita primary energy cons.: | 1.07 toe | 1.15 toe |
| Per capita electricity consumption: | 1,935 kWh | 1,708 kWh |

\* Gross domestic product (GDP) expressed in 1995 prices and PPP.

## BACKGROUND

**Economic development**

By far the largest and most populous country in South America, Brazil accounts for half of the continent's surface area, population and GDP. Brazil's economy ranks ninth in the world. In 2000, it contributed 3% to world GDP, or US$1,200 billion (in 1995 prices and Purchasing Power Parities – PPP). Its 170 million citizens represented a similar share of the world's population. Brazil has a generous endowment of land and mineral resources. By surface area, it ranks amongst the five largest countries in the world. It has large and varied energy resources, particularly hydropower, biomass and hydrocarbons.

Brazil shares a border with all South American countries except Chile and Ecuador, and has an Atlantic Ocean coastline of 7,500 kilometres. With a total area of 8.5 million square kilometres, average population density is low: 20 people per square kilometre. But most Brazilians live along the coast, and 43% of the total population is concentrated in the Southeast region, the country's industrial heartland. This region, which encompasses the states of São Paulo, Rio de Janeiro, Minas Gerais and Espirito Santo, is also the richest part of the country, accounting for over half of Brazil's GDP.[1]

Over the last forty years, developments in the Brazilian economy can be divided into three distinct phases.[2] A period of rapid growth and intense industrialisation driven by strong state intervention in the economy in the 1960s and 1970s was followed by two decades of disruptive cycles of contraction and recovery. The 1980s, the so-called "lost decade", was characterised by featured stagnant income, hyperinflation, high external and public deficit and a series of painful and unsuccessful stabilisation plans. A reorientation of economic policy in the 1990s created the basis for macroeconomic stabilisation and more balanced growth. All in all, GDP increased by an average of 4.7% per year between 1960 and 2000, a very high average over such a long period. However, population also increased rapidly, so that average per capita income increased by only 2.4% per year in the same period.

There are substantial regional differences in terms of level of development: some states (in the Southeast region) have per capita incomes above those found in some European economies, while other states (in the North and Northeast) compare with the poorest countries in the world (see *Table 7.1*). Overall income disparity is partly explained by these regional imbalances, although intra-state inequalities are also significant. Income distribution in Brazil – one of the most skewed in the world – has actually worsened over the last 20 years. Currently, 50% of the country's income is concentrated in the hands of 20% of the population, while the poorest 20% have only 6% of the income.

## The Plano Real and the 1999 financial crisis

In 1994, the federal government launched the *Plano Real*, which introduced widespread structural reforms and accelerated the opening of the Brazilian economy to private investment and international competition. The *Plano Real* kept the Brazilian currency stable. Toward the end of the decade, however, the *real* became seriously overvalued compared to the US dollar, creating insurmountable vulnerabilities for the Brazilian economy in the wake of the Asian and Russian crises of 1997-98.

In January 1999, Brazil had its own monetary crisis, forcing the government to first devalue and then to float the currency, which devalued by about 50%. The effect of the devaluation on the external trade balance, plus sound fiscal and monetary policies

---

1. *Map 10* at the end of the book shows Brazil's regions and states.
2. OECD (2001).

allowed for a quick recovery in 2000. GDP was up 4.5% in 2000 and was predicted to grow by about 4% in 2001. However, the electricity crisis (see *Box 7.1* below) and the global economic downturn following the September 11 attacks in the United States slowed economic growth to 1.5% in 2001. Growth in 2002 has recovered somewhat but, according to Economist Intelligence Unit (EIU) estimates, will not exceed 2-2.5% in 2002.[3]

## Economic reforms and privatisation

The wave of privatisation of public assets reached Brazil later than most of the rest of South America. A far-reaching privatisation programme, the *Programa Nacional de Desestatização* (PND), was launched in 1990, establishing a clear and transparent legal and regulatory framework for divestiture of state holdings. However, it was only with the *Plan Real* in 1994 that the government started to give privatisation high priority. The central development bank, BNDES, was given a central co-ordinating role, which shielded privatisation from excessive political pressure. Overall, the process has been carried out transparently, but the web of ownership ties linking federal government and state authorities has complicated and delayed decisions. This has been a problem particularly for the electricity sector, where market reform and restructuring were not sufficiently addressed before privatisation.

As legal monopolies existed and national ownership was prescribed by law in a number of sectors, including electricity and petroleum, it was necessary to amend the Constitution in order to liberalise ownership of assets. In 1995, constitutional revision 7/95 modified Article 177 of the Federal Constitution to allow private investment in the oil and gas industries.

## Foreign investment

As with other developing countries, Brazil's economic growth potential is linked to the availability of external financing, because domestic savings are typically below the level of investment needed for rapid growth. Throughout the last three decades Brazil has been a major recipient of both external credit and foreign direct investment. Brazil is the second-largest debtor of the multilateral financial organisations and has the largest debt from private sources (banks and bond issues) of all developing countries.[4]

Brazil is also a leading recipient of net foreign direct investment (FDI). These capital inflows are reflected in the important role of foreign companies in the enterprise sector. Foreign companies control about 11% of the capital in Brazil's economy and are responsible for 14% of production.[5] The share of foreign capital in the energy sector soared in the last five years. Currently, foreign investors own 26% of electricity generation and 64% of distribution (compared with 0.3% and 2.3% respectively in 1995). Foreign investors also control the largest gas distribution companies and acquired significant exploration acreage after Petrobras's monopoly over the upstream sector was eliminated in 1998.

Foreign investment is governed by the 1962 Foreign Capital Law and subsequent amendments. In 1995, the Brazilian Congress approved amendments to the Constitution that eliminated the distinctions between foreign and domestic capital and allowed

3. EIU (2002).
4. OECD (2001).
5. CEPAL (2001).

private investment in a number of sectors previously reserved for state companies. The only energy sub-sector where foreign investment is still banned is nuclear energy. Foreign investment in the Brazilian stock exchange has been permitted since 1991.

**Regional integration**

Brazil is a member of the *Mercado Común del Sur* (Mercosur, or Mercosul in Portuguese), a trade block formed in 1991 by Argentina, Brazil, Paraguay and Uruguay, with Bolivia and Chile as associate members. Mercosur was initially intended to be a free-trade zone, then a customs union, and ultimately a common market. However, market integration is still far from complete. Important markets, like sugar, remain partly or fully closed, and common sets of rules are lacking in key policy areas.

Mercosur has contributed to significantly increased intra-regional trade and has helped attract substantial foreign investment in the region. However, since 1999 the macroeconomic and financial difficulties experienced by Mercosur's two major partners have contributed to reduce economic activity and trade flows. Moreover, unilateral decisions taken by member countries have at times resulted in strained relations.[6] In January 1999, the Brazilian currency crisis had an adverse affect on Argentine exports, prompting Argentina to impose import controls in July 1999, which generated much indignation in Brazil. Further difficulties in Mercosur can be expected following the 2002 devaluation of the peso, the imposition of exchange and capital controls and the deep recession in Argentina.

## OVERVIEW OF THE BRAZILIAN ENERGY SECTOR

**A country rich in energy resources**

Brazil has large indigenous energy resources, ranging from abundant hydropower, to plentiful biomass, hydrocarbons, as well as some uranium and coal. At the end of 2001, Brazil's proven oil reserves were estimated at 8.5 billion barrels (1.2 billion tonnes), second only to Venezuela in South America.[7] The country also has significant potential for renewables, such as small hydro, solar and wind power.

Reflecting the size of its economy and population, Brazil's energy consumption is the largest in South America. In 2000, Brazil's total primary energy supply (TPES) amounted to 183 Mtoe[8], equivalent to 46% of total South American primary energy supply. With TPES just above that of Italy or Korea, Brazil is the world's tenth largest energy consumer and the fourth largest non-IEA energy consumer after China, Russia and India.

Oil, biomass and hydroelectricity dominate Brazilian primary energy supply (*Figure 7.1*). Oil and biomass provide about a half and a quarter of TPES, respectively. Hydroelectricity covers on average 90% of the country's electricity needs.

---

6.  Almeida (2001).
7.  Proven reserves of crude oil and natural gas liquids at the end of 2001, according to BP (2002).
8.  Data for energy demand and supply to 2000 in this chapter are taken from the IEA databases (published yearly in *Energy Balances of Non-OECD Countries* and *Energy Statistics of Non-OECD Countries*). IEA data are derived from data provided by the Ministry of Mines and Energy (MME), but differ slightly from the data the MME publishes annually in the *Balanço Energético Nacional* because of the use of different conventions, definitions and conversion factors.

Coal, nuclear power and natural gas play a minor role in the Brazilian energy mix. Brazilian coal is of poor quality, and all the coking coal used in the steel industry is imported. Nevertheless, the use of local coal for power generation is increasing. Nuclear has never accounted for more than 1% of total electricity generation. Power production from Brazil's first nuclear power plant, the 657-MW Angra I located near Rio de Janeiro, has been unreliable since the plant came on stream in 1984. A second 1300-MW nuclear unit at the Angra station started commercial operation in 2001, after construction was stopped for seventeen years. Construction of a third 1300-MW unit started in 1981, but its completion is uncertain because of lack of funds and objections to nuclear power. So far, domestically-produced gas has been used mainly in industry and in the oil and gas sector. Gas imports started in 1999 and are expected to be used largely in the power sector.

**Reducing oil dependence is a key priority**

Despite its large energy resource base, Brazil currently consumes more energy than it can produce. In 2000, net energy imports reached 43 Mtoe: an energy import dependence ratio of 24%. Net imports of crude oil and petroleum products accounted for 23 Mtoe, equivalent to an oil dependence ratio of 31%. In 2001, Brazil spent US$4.6 billion on net imports of crude oil and petroleum products, down from US$6.5 billion in 2000, but higher than in 1999 (US$4.3 billion).[9] Reducing dependence on oil imports has long being a political priority in Brazil. The oil dependence ratio was as high as 90% in 1979, when international oil prices tripled.

**Figure 7.1** Primary energy supply in Brazil by fuel, 1975-2000

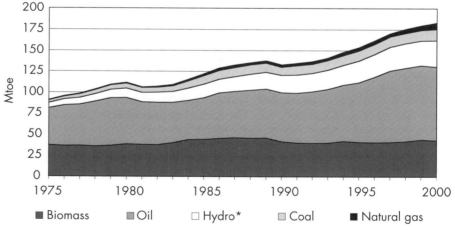

* Includes nuclear and net imports of electricity.
Source: IEA.

**Large use of "commercial" biomass**

Brazil makes extensive use of its biomass resources. Unlike most developing countries, Brazil uses biomass mainly in "modern" or "commercial" applications, rather than in "traditional" or "non-commercial" ones.[10] Indeed, biomass accounts for one-third of the energy used in industry, 14% of the energy used in road transport and 13% of the electricity generated by autoproducers. Sugar cane products play a major role: sugar

9.   ANP's statistical data (http://www.anp.gov.br/petro/dados_estatisticos.asp).
10. See definitions of modern and traditional biomass in the Glossary. For more details on biomass use in developing countries, see IEA (1998c) and IEA (1998d).

cane-based ethanol is used in the transport sector and bagasse, the residual product from sugar cane processing, is used for power generation and is major fuel in the food and beverage industries. In addition, the steel industry uses a substantial amount of charcoal. Charcoal is produced very efficiently in Brazil and much of it comes from sustainable forestry projects. Although the use of charcoal in the steel industry is falling, it was still used in the same quantity as coking coal in 2000. Firewood and charcoal are also important fuels in the residential and agriculture sectors, although their use is decreasing.[11] Taking into account hydropower and biomass, no other large country relies on modern renewable energy to such a degree.

**A unique hydro-dominated power-generation system**

Brazil is the second-largest producer of hydroelectricity in the world after Canada. In 2000, its hydropower capacity of 60.7 GW generated 305 TWh, equivalent to 87% of total electricity generation.[12] *Table 7.1* shows that only the US, China and Canada have larger hydropower capacities than Brazil, but in all three of those countries hydropower accounts for a lesser share of both total capacity and total generation. Among large countries, only Norway has a higher share of hydro in electricity generation (99%) than Brazil.

The uniqueness of the Brazilian power system lies in the fact that the hydro system is linked to a thermal power system which is much smaller than the hydro system. In the US, the hydro system is linked to a much larger thermal system. In Canada, it is linked to a thermal system of roughly equal size. Even the Norwegian system is linked, directly and via Sweden, to the large Western European synchronised system, which is predominantly thermal.

The other particularity is that in Brazil the distances between the various parts of the hydro system (and between the Brazilian hydro system and a neighbouring country's

**Table 7.1** Share of hydropower: international comparisons, 2000

| | Hydropower generation | | | Hydropower installed capacity | |
|---|---|---|---|---|---|
| | TWh | Share of total power generation | | GW | Share of total installed capacity |
| Canada | 358 | 59% | United States | 98 | 12% |
| **Brazil** | **305** | **87%** | China | 79 | 25% |
| United States | 248 | 6% | Canada | 67 | 61% |
| China | 222 | 16% | **Brazil** | **61** | **82%** |
| Russia | 164 | 19% | Japan | 46 | 18% |
| Norway | 142 | 99% | Russia | 44 | 21% |
| Japan | 87 | 8% | Norway | 28 | 99% |

*Source: IEA.*

---

11. Most of the biomass used in the residential and agriculture sector is probably "traditional" biomass.
12. This share is even higher, considering that Brazil takes almost all the output of the binational 12-GW Itaipu power station it shares with Paraguay. Paraguay exports most of its half of Itaipu's output to Brazil. In 2000, Brazil imported 44.2 GW from Paraguay.

thermal power system, such as Argentina's) are very large. Brazil's electricity system is now completely interconnected, except for about 400 small isolated systems, located mainly in the Amazon region, which account for only 2% of total electricity generation. However, the links between the transmission grids of the Northern, Northeast, Southern and Southeast-Centre-West systems are still rather weak.[13]

The high-voltage transmission system plays a major role in optimising hydropower generation from the different hydrological basins and allows the hydropower system to produce up to 30% more output than if the different systems were run on a stand-alone basis. Due to the large water-storage volume of some of the dams, the hydropower system can regulate pluri-annual stocks for a period of about five years (depending on average hydrology).

Brazil has developed a very sophisticated domestic hydropower industry capable of handling all parts of the hydropower system – from construction to co-ordinated operation. It also has substantial experience in dealing with an ecologically-sensitive environment. The cost of hydro capacity so far has been less than US$1,000/kW.

The country still has large untapped hydro resources. The remaining hydroelectric potential measured in terms of firm energy[14] is estimated at 190 GW. However, the hydro resources in the South and Southeast are largely exploited, and most of the remaining reserves are located in the North, some 3,000 kilometres from the industrial and population centres. Environmental considerations related to the flooding of large areas, as well as the cost and challenge of building long transmission lines through dense jungle, will likely prevent construction of new very large hydropower stations.[15] Nevertheless, an increase in hydropower capacity is expected to come from the upgrading of existing large hydropower stations, from the construction of new medium-sized hydropower plants (30-200 MW) and from the reactivation or construction of small ones (up to 30 MW). These options can more easily satisfy environmental concerns and will more easily attract private investors.

Despite this significant remaining hydropower potential, for a number of years the Brazilian government has been calling for an increase in thermal capacity in order to diversify the country's electricity generation mix. The 2001 electricity crisis (see *Box 7.1*) exposed the vulnerability that comes with high dependence on hydropower and gave further impetus to an increase in thermal generation. In 2000, thermal plants, accounted for 10.4 GW, equivalent to 14% of total capacity, and generated 11% of total power generation. Nuclear power accounted for 2 GW and generated less than 2% of total power production in 2000. Thermal capacity is fuelled mainly with oil products and with coal (4.3 and 3.0 Mtoe respectively in 2000). Gas use in the power sector is just starting. In 2000, gas-fired power stations (in large part industrial autogenerators) generated 2.4 TWh (less than 1% of total power generation) and used less than 700 mcm of natural gas. This figure nearly triple in 2001, due to the commissioning of some 1.3 GW of new gas-fired power plants during the year.

---

13. Brazil is successfully operating a high-voltage (750 kV) direct-current line from Itaipu to São Paolo, a technology which might economically be used to transport power over long distances from the North to the Southeast.
14. Defined in Brazilian statistics as the maximum power that can be generated during the worst hydrological period.
15. Apart from the planned 11-GW Belo Monte hydroelectric project on the Xingu River in Pará state in the Amazone, which is due be tendered in 2003.

⊣ Additional thermal capacity will likely be fuelled mainly by gas. Already some 2.7-2.9 GW of new gas-fired power plants were expected to be operational by the end of 2002, as a result of the government's incentives to promote the construction of new gas-fired power plants. However, the introduction of gas-fired capacity in a hydropower-dominated system, such as Brazil's, poses unique challenges, which are discussed in a separate section at the end of this chapter.

**Box 7.1**     The 2001 electricity crisis in Brazil

In 2001, Brazil suffered a severe electricity shortage, brought on by a combination of the worst drought in 70 years and chronic under-investment in electricity generation and transmission capacity in the previous 15 years. Several years of below-average rainfall had left reservoirs in the Southeast and Central-West regions – which together account for 49% of Brazil's population and 64% of its electricity consumption – at 32-33% of capacity. Normally at the start of the dry season they should be at about 50% capacity to avoid the risk of blackouts. The Northeast was also severely affected.

To avoid rolling blackouts, the government implemented a widespread electricity-rationing programme, a measure unknown in the country for 40 years. The programme aimed to reduce electricity consumption by an average of 20% compared to the same period of the previous year, with different requirements for different customers and geographical areas. Launched in June 2001, the programme initially applied to the Southeast and Centre-West and Northeast regions, but it was later extended to the North.

The rationing programme was very successful. The increase in electricity tariffs played a major role in restraining demand: tariffs increased by 50% for users consuming between 200 and 500 kWh per month and by 200% for consumption above 500 kWh per month. Consumers showed great co-operation and responsibility throughout the crisis, quickly adopting energy-saving behaviour, which appears to have lasted longer than initially anticipated. Electricity demand at the beginning of 2002 was about 15% lower than at the beginning of 2001, but in the second half of 2002 demand picked up again and reached almost pre-rationing levels.

While the electricity shortage was precipitated by an exceptionally dry summer in 2000-2001, the tightness of supply had been growing progressively more acute in the last few years, with electricity demand growth far outstripping the pace of expansion of power generation capacity. During the 1990s, electricity consumption grew at an average of 6.7% per year, while installed capacity increased by only 5.4% annually, gradually eroding generation reserve capacity. In the last few years, the system was operating much too close to full capacity, and reservoirs were drawn down. Insufficient investment in transmission also played a role in the crisis. The South was largely unaffected by the drought, but limited transmission capacity precluded larger electricity exports from the Southern states (and *a fortiori* from Argentina and Paraguay) to the rest of Brazil.

The crisis was declared over at the beginning of 2002, and the rationing measures were withdrawn on 28 February 2002. Above-average rainfalls have now replenished the reservoirs to normal levels in most of the affected states, and demand has been slow to catch up. These two factors, together with the commissioning of a number of gas-fired power stations under the Thermoelectricity Priority Programme and the Emergency Programme, have resulted in oversupply, at least for the short term[16].

Nevertheless, Brazil still needs substantial new capacity to ensure reliable electricity supply in the medium and long term. The government estimates that 46 GW of new plant (both gas-fired and hydropower), as well as 1 GW of additional imports from Argentina and 47,000 kilometres of new transmission lines, will be required in the next ten years to supply demand expected to grow by 5.3% per year[17]. The total investment needed for generation and transmission is estimated to be at least US$50 billion, most of which, the government hopes, will come from the private sector. This, however, will require adequate electricity pricing and a clarification of the structure and regulation of the electricity sector.

## Large disparities in per capita energy and electricity consumption

Economic disparities are reflected in per capita levels of energy consumption (*Table 7.2*). Despite rapid growth in total energy consumption, Brazil's per capita energy use is still comparatively low. The average Brazilian consumes twice as much energy as the average Indian and 20% more than the average Chinese, but only 60% as much as the average Argentine and 30% as much as the average European.[18]

Internally, there are great differences between the rich and populated South and Southeast, and the poorer North and Northeast: the average levels of per capita electricity consumption can differ as much as 6:1 between the richest and poorest states. Access to electricity and modern fuels is still limited in remote rural areas because of the lack of grids and delivery networks. Government surveys indicate that some 100,000 communities and over three million rural properties, representing approximately 20 million people (12% of total population), still have no access to electricity.

## Energy-sector reforms and institutions

Since the mid-1990s, Brazil has implemented significant reforms in the energy sector, starting with a modification of the Constitution in 1995 that allowed private participation in the electricity and hydrocarbon sectors, industries that were previously reserved to the state.

In the hydrocarbon sector, a new licensing regime was instituted, with competitive bidding and no restrictions on nationality. The first bidding round was held in June 1999 and there has been one every following year. Petrobras itself is being restructured and the government has reduced its stake in the company to the compulsory 50% +1. Power-sector reform has been more problematic, as described in *Box 7.2*.

---

16. The government even expects slight overcapacity in 2004-2005.
17. CCPE (2001), Plano Decenal de Expansão 2001-2010.
18. Of course, climatic conditions account for part of the difference: there is no need for space heating in Brazil, while space heating in winter months accounts for a substantial share of final energy use in both Argentina and in Europe.

**Table 7.2** Per capita GDP and electricity consumption in Brazil, 2000

|  | Population (2000) million | GDP per capita (2000) R$ | Electricity consumption per capita (1999) kWh |
|---|---|---|---|
| **BRAZIL** | **169.8** | **5,740** | **1,855** |
| **North** | 4.0 | 3,380 | 1,243 |
| Amazonas | 2.8 | 5,577 | 946 |
| Tocantins | 1.2 | 1,832 | 557 |
| **Northeast** | **26.2** | **2,671** | **1,074** |
| Maranhão | 5.7 | 1,402 | 1,426 |
| Piauí | 2.8 | 1,660 | 466 |
| Bahia | 13,1 | 3,206 | 1,265 |
| **Southeast** | **69.3** | **7,843** | **2,522** |
| Minas Gerais | 17.9 | 5,239 | 2,211 |
| Rio de Janeiro | 14.4 | 7,946 | 2,447 |
| São Paulo | 37.0 | 9,210 | 2,678 |
| **South** | **10.2** | **6,878** | **1,950** |
| Rio Grande do Sul | 10.2 | 7,389 | 1,895 |
| **Middle-West** | **4.6** | **5,421** | **1,382** |
| Mato Grosso | 2.5 | 4,695 | 1,181 |
| Distrito Federal | 2.1 | 10,935 | 1,766 |

*Sources: IBGE website, MME.*

**Box 7.2**          Power-sector reform in Brazil

When the current administration took office in 1995, 23 large power plants, with a combined capacity of 10 GW (one-fifth of the system's total capacity) lay unfinished for lack of funds. Another 33 projects had not been started. The Brazilian electricity sector had traditionally been expanded through external financing, government resources and electricity tariffs. In the early 1980s, international lending dried up due to the debt crisis; then fiscal austerity reduced government spending on infrastructure. To make matters worse, electricity tariffs were not allowed to rise in line with inflation, further undercutting investment by federal and state-owned electricity utilities and leaving the whole sector with huge debts. In 1993, with a debt of around US$20 billion, the Brazilian electricity sector was so close to bankruptcy that the federal government had to intervene. Reform was inevitable.

The new administration embarked on a major power sector reform including tariff adjustment and financial rehabilitation of the electricity utilities, the unbundling of vertically-integrated utilities, the divestiture of public assets, the creation of an independent non-profit national system operator, ONS, the definition of rules for a wholesale energy market, MAE and the establishment of the electricity regulatory agency, ANEEL. Yet the transition from a centralised public system to a competitive, market-driven, private system proved more difficult than expected.

After a promising start (26% of the generation assets and nearly 70% of the distribution were privatised by the end of 1999, generating revenues of US$21 billion), the reform stalled due to political resistance to further unbundling and privatisation of federally-owned regional utilities. One major problem was that, in contrast to best practice, privatisation of public assets began before the norms for the new power sector were fully defined and before the institutions that would apply them were fully operational.

The Brazilian electricity system thus remained in the dangerous position of no longer being centrally planned, but lacking the proper incentives and market signals for companies to invest in much-needed new infrastructure. Uncertainty about the future institutional and regulatory framework contributed to delaying investment by public companies[19], while scaring off private companies. Private investors interested in the Brazilian electricity sector opted to buy distribution companies and existing power plants, a much less risky investment (and more lucrative in the short term) than greenfield projects.

In addition, the 1999 financial crisis and currency devaluation renewed concerns about currency risk, an important consideration in an industry where many of the investments are made in US dollars while sales are in local currency and tariffs are capped. As for gas-fired power projects, additional factors, such as tightness in the global gas-turbine market, difficulties in obtaining the necessary environmental permits and a lack of trained manpower to operate the planned power stations, contributed to delaying projects.

Between 1995 and 2000, the unfinished projects from the previous period were concluded. In total, 16 GW of new capacity were added to the system, as well as more than 7,000 kilometres of transmission lines. However, these additions were not sufficient to compensate for 15 years of under-investment and electricity demand growing at nearly 5% per year. Even before the full-blown power crisis in 2001, brownouts and blackouts were common.

The 2001 power crisis had at least one positive consequence: it forced electricity and energy policy reform to the top of the government agenda. A inter-ministerial Committee for the Revitalisation of the Electricity Sector was set up in early 2002 and put forward a number of measures to solve some of the problems that had stalled the reformed privatisation of the electricity sector. The proposals are currently undergoing public consultation and are expected to be adopted by the end of 2002.

The Ministry of Mines and Energy (MME), created in 1960, is the federal government body responsible for the energy sector. Within this ministry, energy issues are dealt with by the Secretariat of Energy, which is subdivided into two departments: the National Department for Energy Policy and the National Department for Energy Development.

19. The IMF, which has lent Brazil billions of dollars in the past few years to avoid a financial meltdown, insisted that Brazilian state companies should not make new investments, saying that these should be left to the private companies which eventually bought the assets. But with few assets actually sold, no investments at all have been made.

In 1997, a new inter-ministerial body was created, the National Council for Energy Policy (*Conselho Nacional de Política Energética* – CNPE). The CNPE is headed by the Minister of Energy and Mines and reports directly to the President. The CNPE should oversee the entire energy sector and issue national energy policy directives, which the MME elaborates and implements, defining objectives and means. However, the CNPE met for the first time in December 2000 and, until the electricity crisis, it did not play a major role.

The National Petroleum Agency (*Agência Nacional de Petróleo* – ANP) is the regulatory agency for the oil and gas sector. Created in 1997, it is an autonomous agency of the Federal Public Administration, linked to the MME. The National Electrical Energy Agency (*Agência Nacional de Energia Elétrica* – ANEEL), created in 1996, is the regulatory agency for the electricity sector.

Because of the federal structure of the country, local regulatory agencies were created in some states. Responsibility for the gas sector in particular is split between the federal and state governments: the authority of the federal government extends from exploration and production to transportation up to the city-gates, while state governments are responsible for natural gas distribution activities. Regulatory authority for downstream gas belongs to the State Energy Secretariats, some of which have created regulatory agencies to oversee and regulate the gas sector. Apart from the State of São Paulo, which has a specific regulatory agency for electricity and gas (CSPE), the other state regulatory agencies are all multi-sectorial, with responsibilities for electricity, gas, telecommunications, transport, sanitation and water. Of recent creation, and often lacking personnel and expertise, the state regulators have so far played a limited role in the gas sector.[20]

Two ministries whose responsibilities overlap in certain areas with that of the MME are the Ministry of Science and Technology (MCT) and the Ministry of the Environment (MMA). Both are represented in the CNPE. In addition, the National Development Bank (BNDES) is responsible for overseeing Brazil's privatisation process, including the sell-off of gas distribution companies. The BNDES also has an important role in the financing of new energy projects.

## NATURAL GAS RESERVES AND PRODUCTION

Brazil has modest proven gas reserves accounting for just 3% of the region's total proven reserves. According to the ANP, as of 1 January 2002, Brazilian proven gas reserves stood at 220 billion cubic metres (bcm). Total proven and probable reserves are estimated at 332 bcm. Despite this low level of reserves, Brazil is thought to have substantial potential for new gas discoveries. The 2000 US Geological Survey (USGS) estimated Brazil's undiscovered natural gas resources at over 5 trillion cubic metres (tcm), about

---

20. Apart from CSPE, thirteen states have regulatory agencies that deal with natural gas, including Rio de Janeiro (ASEP), Rio Grande do Sul (AGERGS), Bahia (AGERBA), Ceará (ARCE), Pará (ARCON) and Rio Grande do Norte (ARSEP).

40% of total estimated undiscovered gas resources for the whole South American continent (see *Table 1.4* in *Chapter 1*).

The largest gas fields are in the offshore Campos and Santos basins, off the coast of the state of Rio de Janeiro and in the Urucu Basin in the Amazon forest. As a result, the state of Rio de Janeiro in the Southeast accounts for 48% of the country's natural gas reserves and for 42% of natural gas production (*Table 7.3*). The state of Amazonas in the North has the country's second-largest reserves (20% of total), but production is for now entirely reinjected after the extraction of NGL because of long distances and difficult access to potential markets. Reserves in the Northeast, largely in the states of Bahia and Rio Grande do Norte, account for another 18% of total reserves.

In 2001, Brazil's gross natural gas production reached 14 bcm, up 5.4% from the 13.3 bcm produced in 2000. Production of natural gas in Brazil started in 1954 in the state of Bahia. Only in the mid-1970s did gas production develop in other neighbouring states in the Northeast. In 1980, Bahia produced 3 million cubic meters per day (mcm/d) accounting for half of the country's production. Gas production in the offshore Campos basin started around 1980, growing rapidly to 6 mcm/d by the mid-1980s and over 15 mcm/d in 2000. The breakdown of current production by states and regions is shown in *Table 7.3*.

Because many of the natural gas reserves are associated with oil and in offshore fields, a large portion of natural gas production has traditionally been reinjected or flared (*Figure 7.2*). Currently about 60% of gas production is from offshore wells and 80%  is associated with oil. The Bahian fields (largely onshore) account for 45% of non-

**Table 7.3** Natural gas reserves and production by State in Brazil, 2001

| | Reserves, at 1.01.2002 | | Gross production 2001 | |
| | Proven | Probable | | |
| | bcm | bcm | mcm | % |
|---|---|---|---|---|
| **North:** | **44,549** | **75,324** | **2,427** | **17%** |
| Amazonas | 44,549 | 75,324 | 2,427 | 17% |
| **Northeast:** | **52,917** | **71,594** | **4,832** | **34%** |
| Ceará | 1,186 | 1,239 | 93 | 1% |
| Rio Gr. do Norte | 19,848 | 19,223 | 1,198 | 9% |
| Alagoas | 6,920 | 10,155 | 763 | 5% |
| Sergipe | 4,996 | 7,374 | 812 | 6% |
| Bahia | 19,967 | 33,603 | 1,966 | 14% |
| **Southeast:** | **122,306** | **182,928** | **6,701** | **48%** |
| Espirito Santo | 11,787 | 19,230 | 389 | 3% |
| Rio de Janeiro | 106,246 | 159,425 | 5,968 | 42% |
| Sao Paulo | 4,273 | 4,273 | 344 | 2% |
| **South:** | **68** | **2,527** | **85** | **1%** |
| Paraná | 68 | 2,527 | 85 | 1% |
| **Total Brazil:** | **219,840** | **332,373** | **14,045** | **100%** |

*Source: ANP.*

associated gas production, while the offshore Campos fields in the state of Rio de Janeiro account for over 50% of associated-gas production.

According to ANP data, 22% of total gross natural gas production in 2001 was reinjected, 19% was flared or vented and 7% was lost during NGL extraction (volume shrinkage). Hence, just over 50% of gross natural gas production was available for commercialisation (marketed production)[21]. As indicated in *Figure 7.2*, the proportion of gas flared in Brazil has fallen over time, although it increased in absolute terms and now exceeds 2 bcm/year. Petrobras recently launched a "Programme of Zero Flaring". Over the last 30 years, marketed gas production grew by 15% per year, well above the growth rate for gross production during the same period (8% per year).

**Figure 7.2** Natural gas gross and marketed production, 1975-2001

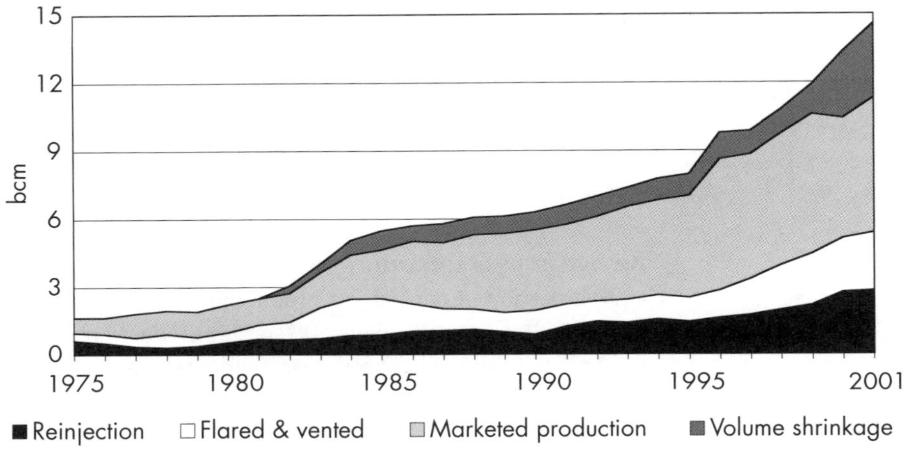

■ Reinjection   □ Flared & vented   ▨ Marketed production   ■ Volume shrinkage

*Source: Cedigaz.*

**Petrobras is the dominant gas supplier**

*Petróleo Brasileiro S.A.* (Petrobras), the Brazilian state-owned oil company, is currently the only natural gas producer in the country. It also owns substantial gas reserves and production in Bolivia, and has preferential access to the Bolivia-Brazil *Gasbol* pipeline. Petrobras is also the dominant gas importer. Hence, Petrobras currently supplies virtually all the gas in the country. The company is also expanding its interests into gas transportation and distribution, as well as in gas-fired thermal generation. It owns virtually 100% of all national pipelines through its subsidiary Transpetro and has a 51% stake in the Bolivia-Brazil pipeline's operator (TBG). It also has stakes in 17 of the 22 existing gas distribution companies, and in 18 of the 44 thermal power plants currently under construction under the Thermal Power Priority Programme.

Petrobras is the largest company in Brazil. Its core business includes exploration and production, refining, processing, transportation and commercialisation of crude oil, oil products and natural gas. In response to the new competitive environment, Petrobras has successfully undergone thorough internal restructuring and has taken steps to

---

21. This calculation is consistent with the definition used in IEA and others' statistics and balances and adopted in this report (see the annex on Definitions at the end of the report). In reality, however, the amount of gas supplied to the market is even less since another 12% of gross production is used directly by the E&P sector, mostly as a fuel for offshore platforms.

transform itself from a monopolistic national oil company into an efficient, diversified and international energy company.

Its new organisational structure, implemented in November 2000, groups some 40 business units into four business areas – Exploration & Production (upstream), Supply (downstream), Gas & Energy and International – as well as two support areas, Finance and Services. In addition to its holding operations, Petrobras has five independent subsidiaries, each with its own board of directors, but linked to the head office:

■ *Petrobras Gás S.A.* (Gaspetro), responsible for retailing Brazilian and imported natural gas;

■ *Petrobras Química S.A.* (Petroquisa), which operates in the petrochemical sector;

■ *Petrobras Distribuidora S.A.* (BR Distribuidora), responsible for the commercialisation of oil products;

■ *Petrobras Transporte S.A.* (Transpetro), created in 1998 to comply with the 1997 Oil Law to take over the operation of Petrobras's offshore and onshore oil and gas pipeline transportation and the management of oil, by-product and natural gas terminals;

■ *Petrobras Internacional S.A.* (Braspetro), responsible for Petrobras's overseas operations in the different segments of the oil and gas industry. Independently or through its subsidiaries and branches offices, or in partnership with the world's major oil corporations, Braspetro is present on four continents and in 32 countries, including Angola, Argentina, Bolivia, Colombia, Equatorial Guinea, Nigeria, Trinidad & Tobago and the United States. Braspetro is involved in the upstream sector (exploration & production), in downstream activities (refining, marketing and sales, transportation and logistics), as well as in the provision of engineering and oil-well drilling services.

In July 2000, the government sold 30% of Petrobras's shares on the national and international capital markets. In 2001, the BNDES sold 41,381 preferred shares of Petrobras, representing 3.5% of the company's total capital, which it has managed for the federal treasury. As a result, the federal government now controls 55.7% of the voting shares or 32.5% of total capital. Total privatisation of Petrobras is barred by the Petroleum Law, which stipulates that the government must retain a majority of the voting shares, and would likely to be vehemently opposed by the Congress and the general public. However, the law authorises Petrobras to create affiliates in which it may be a minor shareholder.

**Other potential gas producers**

Since the approval of the 1997 Petroleum Law, Petrobras no longer has an exclusive monopoly over exploration and production of oil and natural gas. Forty-nine companies, including Petrobras, have acquired rights to explore and produce oil and gas in Brazil in the four E&P bidding rounds held by the ANP since 1998. These companies include all the majors, numerous middle and small foreign companies and a number of Brazilian companies. New entrants are actively exploring, and some gas discoveries have already been reported, but Petrobras is expected to remain the dominant player at least for the coming decade. other players in gas

## NATURAL GAS IMPORTS

Since 1 July 1999, Brazil has been importing increasing gas volumes from Bolivia through the *Gasbol* pipeline. The Gasbol pipeline has been a major factor in increasing Brazil's natural gas supply. When operating at full capacity,[22] it will be able to transport 30 mcm/d of natural gas, providing about half of Brazil's projected natural gas requirements. São Paulo State alone is expected to absorb half of the capacity of the pipeline.

In addition, imports of Argentine gas started in July 2000. Currently, the *Paraná-Uruguayana* pipeline supplies a 600-MW power station located at Uruguaiana, near the Argentina-Brazil border, in the State of Rio Grande do Sul (Southern region). This pipeline is the first portion of a more ambitious project to supply Argentine gas to the city of Porto Alegre in southern Brazil. When completed, the *Paraná-Porto Alegre* pipeline will have a capacity of 12 mcm/d.

In 2001, Brazil imported 4.6 mcm of natural gas, twice as much as in 2000, at a cost of US$365 million. This corresponds to an average price of US$79/1,000 m$^3$ (equivalent to US$2.12/MBtu, at a calorific value of 9,400 kcal/m$^3$). Of total imports in 2001, 84% came from Bolivia and the rest from Argentina. *Figure 7.3* illustrates the recent trends of Brazil's natural gas imports.

**Figure 7.3** Brazilian natural gas imports by company, 1999-2002

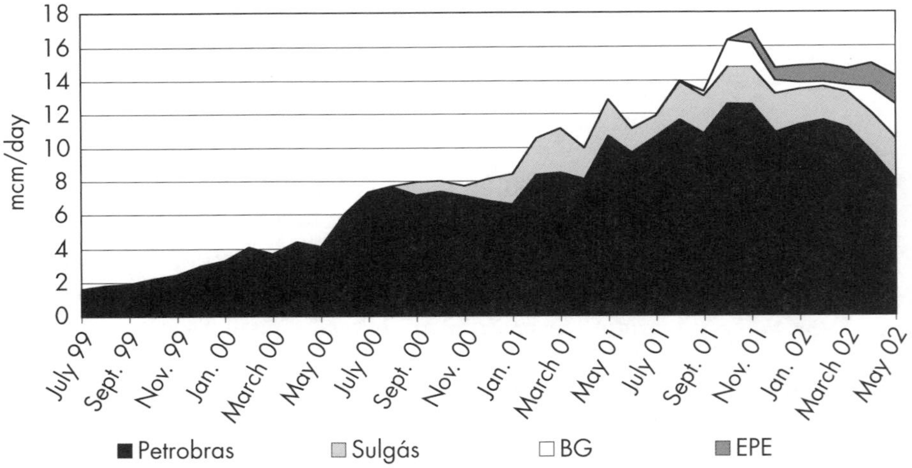

Source: ANP.

**The Bolivia-Brazil (Gasbol) pipeline**

Since mid-1999, Brazil has been importing gas from Bolivia through the Gasbol pipeline. The pipeline extends 2,593 kilometres in Brazil and 557 kilometres in Bolivia, making it the largest gas pipeline in Latin America (see *Map 11* at the end of the book). It passes by 135 Brazilian towns in five states (Mato Grosso do Sul, São Paulo, Paraná, Santa Catarina and Rio Grande do Sul) and indirectly benefits two others states (Rio de Janeiro and Minas Gerais).

---

22. Full capacity is expected to be reached in May 2003, when all compression stations are in operation.

The pipeline was built in two stages at a total cost of US$2.1 billion. The first stretch of the pipeline – 1,418 kilometres long with a diameter varying from 32 inches to 24 inches – was inaugurated in February 1999 and started operation on 1 July 1999. It begins at Rio Grande in Bolivia, crosses into Brazil at Corumbá in the state of Mato Grosso do Sul and reaches Campinas in the state of São Paulo. The second stretch, linking Campinas in São Paulo State to Canoas in Rio Grande do Sul, was completed in March 2000. It is 1,165 kilometres long with a diameter varying from 24 inches to 16 inches.

The Bolivia-Brazil pipeline is the single largest private-sector investment in South America. The idea for natural gas trade between Bolivia and Brazil had been around since the 1930s.[23] In 1990, the two governments decided to give the gas export pipeline another serious look and in 1993 the two (then) state monopolies, Petrobras in Brazil and Yacimentos Petroliferos Fiscales Bolivianos (YPFB) in Bolivia, signed a gas sales contract. When the two countries decided to proceed with the project in the early 1990s, it was clear that neither country could afford to finance it from the public budget. The debt crisis of the 1980s had severely reduced the ability of governments and public companies to borrow on the international markets, and there also was a growing perception that large infrastructure projects were best left to the private sector. The successful privatisation of energy assets in nearby Chile and Argentina and the rapid increase in private capital flows to developing countries reinforced this perception. So both Brazil and Bolivia embarked on efforts to find private equity partners for a pipeline company on each side of the border. The resulting equity structure on both sides is shown in *Table 7.4*. The project structure comprises a certain degree of cross-ownership by each sponsor group and special committees were formed with representation from all sponsors.

**Table 7.4** Current ownership structure of Gasbol's transportation companies

| Company | Stake | Constituents |
|---|---|---|
| **Gas Transboliviano, GTB** | | |
| **(Bolivian gas transport company)** | | |
| **Bolt JV** | 85% | Shell/Enron: 40% <br> Transredes (a 50/50 partnership of Shell/Enron and Bolivian Pension Funds): 60% |
| **BBPP Holdings** | 6% | TotalFinaElf: 33.3% (*) <br> El Paso Energy: 33.3% <br> British Gas: 33.3% |
| **Gaspetro** | 9% | Petrobras: 100% |
| **Transportadora Brasileira Gasoduto Bolivia-Brasil, TBG** | | |
| **(Brazilian gas transport company)** | | |
| **Gaspetro** | 51% | Petrobras: 100% |
| **BBPP Holdings** | 29% | TotalFinaElf: 33.3% (*) <br> El Paso Energy: 33.3% <br> British Gas: 33.3% |
| **Shell** | 4% | |
| **Enron** | 4% | |
| **Transredes** | 12% | |

\* *TotalFinaElf bought its stakes from BHP in 2000.*
*Sources: TBG, GTB.*

---

23. Law and de Franco (1998).

In 1997 the project still lacked a firm financing plan. Market soundings had indicated a lack of capacity for long-term commercial funding. Commercial debt would be high-cost with short maturities (eight to ten years) because of perceived Brazilian country risk, regulatory risk and supply risks resulting in debt-service difficulties and a final cost for the gas that could severely limit successful market penetration. In 1997, the World Bank and its multilateral counterparts decided to appraise the project on the understanding that transmission tariffs would be regulated to ensure that any benefit from their loans and guarantees would be passed on to the final consumer. The World Bank agreed to provide for the Brazilian side a direct loan of US$130 million and a partial credit guarantee of US$180 million to TBG. Other multilateral organisations provided financing totalling US$380 million. The multilateral financing covered 40% of the financing requirement as senior debt; Petrobras provided another 40% sourced from bilateral agencies and equity sponsors provided the rest.

On the Bolivian side, only 20% financing was available from shareholder equity. With the Bolivian government unprepared to provide sovereign guarantees, little progress was made to close the financing gap. Realising that this threatened to delay the project, the Brazilian government urged Petrobras to seek a solution. Petrobras responded through two mechanisms. First, it agreed to finance a fixed-price, turnkey construction contract for the Bolivian section of the pipeline, with repayment through the waiver of future transportation fees. Second, it agreed to pre-purchase part of the uncommitted upside capacity of the pipeline on both sides of the border, an arrangement that became known as the transport capacity option.

**Table 7.5** Financing of the Gasbol pipeline (1997 US$ millions)

| Funding source | GTB (Bolivia) | TBG (Brazil) |
|---|---|---|
| Shareholder equity (including subordinated loans) | 75 | 310 |
| Petrobras transport capacity option, with Brazilian National Development Bank and Andean Development Corporation financing | 81 | 302 |
| Petrobras loan, with Jexim/Marubeni and Brazilian National Development Bank financing | | 348 |
| Petrobras advance payment contract, with Jexim/Marubeni financing | 280 | |
| World Bank loan | | 130 |
| World Bank partial credit guarantee | | 180 |
| Inter-American Development Bank | | 240 |
| Corporación Andina de Fomento | | 80 |
| European Investment Bank | | 60 |
| **Total** | **436** | **1,650** |

*Sources: TBG.*

Petrobras bears most of the project risks on both sides of the border. The biggest risk lies in the market in Brazil. Although the ultimate risk lies with the distribution companies, it is Petrobras that is obliged to pay YPFB for the gas and the transportation companies for their transport services. Moreover, through its turnkey construction contract, Petrobras took the construction risk on the Bolivian side.

The contract between Petrobras and YPFB allows for a gradual build-up of deliveries, in order to allow for the development of the Brazilian gas market and the ramp-up of the pipeline's capacity as compression stations are gradually installed. The current pipeline capacity is around 17 mcm/d. Full capacity of 30 mcm/d (11 bcm/year) is expected by mid-2003, when all the compression stations will be operational.

There are plans to expand the capacity of the pipeline to 50 mcm/d. TBG launched an open-season for the expansion in November 2001. Several companies submitted declarations of interest for a total of 21 mcm/d of new capacity: British Gas, El Paso, Pan American Energy, Petrobras, Repsol, Shell, TotalFinaElf and two Brazilian glass manufacturers, Guardian do Brasil Vidros Planos and Nadir Figueiredo. However, the deadline for the closing of the open season has been postponed several times due to uncertainty about new electricity sector regulations and the impact on future gas demand of planned thermal power plants.

## The Lateral Cuiabá pipeline

A second Bolivia-Brazil pipeline started operations in 2002. This is a 626-km spur off the Gasbol pipeline. It starts at Rio San Miguel in Bolivia, crosses into Brazil at Cáceres and proceeds to Cuiabá in the State of Mato Grosso. In Cuiabá, it fuels a 480-MW thermal-power station. The pipeline (360 kilometres in Bolivia, 266 kilometres in Brazil) has a diameter of 18 inches and a capacity of 2.8 mcm/d. This integrated project was developed by Enron in alliance with Royal Dutch Shell and Transredes at an estimated cost of US$570 million. GasOcidente do Mato Grosso Ltda (Enron 50%, Shell 37.5%, Transredes 12.5%) operates the Brazilian portion of the pipeline, while GasOriente Boliviano Ltda. operates the Bolivian portion. The pipeline will transport gas from Bolivia, but also from Argentina: the Cuiabá power station will receive 2.2 mcm/d from Argentina.

## The Paraná-Uruguaiana Porto Alegre pipeline

Since July 2000, Brazil has also been importing gas from Argentina to supply a 600-MW thermoelectric power station at Uruguaiana, a town in the state of Rio Grande do Sul on the border with Argentina. This is the first stage of a more ambitious project to supply Argentine gas to Porto Alegre in southern Brazil, thereby completing the gas link between Bolivia, Brazil and Argentina.

On the Argentine side, a 440-km, 24-inch pipeline operated by Transportadora de Gas de Mercosur (TGM), takes gas from the Argentine TGN network at Aldea Brasilera (Paraná) to Paso de los Libres on the Argentina-Brazil border. TGM is a joint venture between Compañia General de Combustibles, Techint, TotalFinaElf and Panamerican Energy. The "anchor" project that made the TGM pipeline possible is the Uruguaiana power station, which is owned by the US company AES consumes some 2.8 mcm/d.

The short 25-km stretch in Brazilian territory is operated by Transportadora Sul Brasileira de Gas (TSB), a joint venture between Gaspetro, Petrobras's wholly-owned natural gas subsidiary (25%); French TotalFinaElf (25%); Brazilian Ipiranga (20%), Spanish Repsol-YPF (15%) and Argentine Techint (15%).

When completed, the TSB project will link Uruguaiana to Porto Alegre at an expected total cost of US$350 million. The 615-km, 24-inch pipeline will eventually have a capacity of 12 to 15 mcm/d. Construction of the final 25-km portion from Porto Alegre to the Triunfo petrochemical complex has been completed. Triunfo currently receives gas from Bolivia through the Gasbol pipeline.

Work on the final 565-km middle stretch has been delayed for nearly two years because of uncertainties about the demand side in Brazil and about the supply side in Argentina. In Brazil, uncertainties about the new electricity and gas sector regulations have delayed the construction of two gas-fired power stations near Porto Alegre which would take some 5 mcm/d of the pipeline's capacity. In Argentina, capacity expansion of the TGN network is needed to supply the new line, but the economic crisis has halted all capacity expansion plans.

Once the pipeline is concluded, it will be possible to supply Argentine gas to the São Paulo area by reversing the flow in the southern leg of the Gasbol, thereby increasing supply competition, especially in the gas markets of São Paulo and Rio de Janeiro.

## The Cruz del Sur extension

Several other pipelines to transport Argentine gas to southern and southeastern Brazil's industrial and population centres have been proposed or are at various stages of development. The most advanced is an extension of the *Cruz del Sur* pipeline, which currently connects Buenos Aires with the Uruguayan cities of Colonia and Montevideo. The extension would start in Colonia and extend 415 kilometres in Uruguay and 505 kilometres in Brazil to Porto Alegre. The Cruz del Sur consortium includes Britain's BG with a 40% stake; Pan American Energy (a joint venture between BP and Bridas Energy of Argentina) with 30%; Germany's Wintershall with 10% and the Uruguayan state oil company Ancap with 20%. The US$400-million project aims at supplying 12 mcm/d to the same market targeted by the TSB pipeline.

It is not clear that southern Brazil's gas demand would justify two pipelines – the region has substantial hydropower and also cheap local coal. But the ultimate goal is to supply the much bigger markets of São Paulo and Rio de Janeiro, where Argentine gas could successfully compete with Bolivian gas.

## The Gasin project

In December 2001, the Brazilian government announced plans for another major gas pipeline linking Brazil with Bolivia, Argentina and Paraguay. This so-called *Gas Integration (Gasin)* pipeline would extend 5,250 kilometres in the four countries and cost an estimated US$5 billion. The pipeline would start near the gas fields in southern Bolivia and run through gas-rich northern Argentina, where a spur would cross the border to Asuncion, the capital of Paraguay. The pipeline would then enter Brazil at the western border of Santa Catarina State and run 3,450 kilometres inside Brazil, east and then north all the way to the federal capital Brasilia, supplying a number of large urban agglomerations on the way. Besides the estimated US$5-billion investment for the construction of the pipeline itself, the project is expected to generate up to US$7 billion more in investments for expansion of local gas distribution systems and related industrial co-generation, thermal generation and compressed natural gas (CNG) projects.

The government expects the Gasin pipeline to be financed by the private sector in partnership with state-owned Petrobras with the support of public and multilateral institutions such as the Brazilian Development bank (BNDES), the Andean Development Corporation (CAF), the Inter-American Development Bank, the World Bank, as well as bilateral agencies, private banks, suppliers and trading companies which traditionally finance infrastructure projects. Petrobras and Italy's Snam, of the ENI group, have already signed an agreement to begin economic and technical feasibility studies, together with the Tiete-Parana Development Agency (ADTP) which drew up the basic concept for the project. According to initial plans, the feasibility study is to be concluded in 2002, and the pipeline is to go on stream in 2005.

## NATURAL GAS DEMAND

According to IEA statistics, primary gas demand in Brazil increased from less than 200 mcm in the early 1970s to 9 bcm in 2000 (equivalent to 7.5 Mtoe on a net calorific basis), averaging an annual growth of nearly 16%. Despite its recent rapid growth, the Brazilian gas market is still in its infancy. The share of natural gas in Brazil's energy mix remains very low by regional and international standards. In 2000, natural gas accounted for 4.1% of TPES, 3.1% of total final consumption (TFC) and less than 1% of power generation.

Most of the gas is used in the industrial sector, and a substantial portion is used in the oil and gas sector itself. In 2000, 58% of total gas supply was used by industry, largely in the petrochemical sector; 26% was used by the oil and gas sector; 7% was used to produce electricity; 3.5% was used in transport and 2.2% in the residential commercial and public buildings sector. These shares are evolving rapidly. Preliminary data for 2001 indicate that the share of gas used for power generation reached 15% of total gas supply in 2001, due to the new gas-fired power stations that came on stream. The share of gas used in transport also grew markedly, from 3.5% of total gas supply in 2000 to 5% in 2001.

**Figure 7.4** Natural gas demand in Brazil, 1975-2000

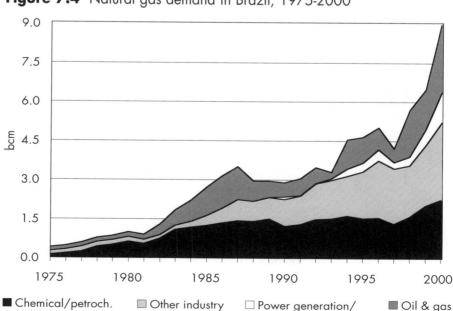

Source: IEA.

## Not one, but three gas markets

Because Brazil is a country the size of a continent, several distinct natural gas markets can be expected to develop, each characterised by its own supply sources, demand centres and transportation network.

Geographically, one can distinguish three natural gas markets in Brazil. The largest and most developed system by far comprises the South, Southeast and Centre-West states. Supplied by imported gas from Bolivia and Argentina and by domestic gas from the offshore basins of Campos, Santos and Espirito Santo on the Atlantic coast, this immense area includes the main population and industrial centres of the country. The region represents 82% of Brazil's industrial output and 66% of the country's GDP. Its 90 million inhabitants consume 75% of the national energy supply. The backbone of the natural gas transportation system of this area is the Bolivia-Brazil *Gasbol* pipeline. Other pipeline trunks link the Campos Basin with Rio de Janeiro and São Paulo.

The coastal cities of the Northeast constitute Brazil's second-largest natural gas system. Now nearly completely interlinked by the Nordestão Pipeline, this area is currently supplied only by domestic gas from the basins offshore the Rio Grande do Norte, Sergipe, Alagoas and Bahia states, though two projects for LNG regasification plants have been proposed.

The third system, with abundant reserves but a market still to be developed, is in the Amazon region. It has not yet been decided if the gas from the Urucú and Juruá fields will be distributed to the region's main centres by pipeline or by barge. At the moment, all the gas extracted is processed to extract valuable liquids and then reinjected. Gas could bring substantial economic benefits to the region by substituting for costly oil products in industry and power generation.

*Table 7.6.* shows the evolution of natural gas sales by region and state. In 2001, 56% of gas demand was concentrated in the Southeast, with the states of Rio de Janeiro and São Paulo accounting for nearly 50% of total demand; 29% is consumed in the Northeast and 14% in the Southern region.

## A slow gas market development...

Several reasons explain the late development of natural gas in Brazil. First, the limited access to gas reserves: Brazilian proven gas reserves are small and before the mid-1990s the combined proven gas reserves of Bolivia and Argentina were thought to be just enough to supply the Argentine market for a decade or two.

This is, however, only part of the story. For many years, the state-owned oil company Petrobras (like many other oil companies around the globe) paid little attention to gas, preferring to concentrate on the production and sale of crude oil and products. Associated gas was considered merely a by-product, useful to boost oil production through reinjection and convenient to produce heat and electricity on offshore platforms. The state oil company did build a few gas processing plants and pipelines to extract valuable NGL and sell the remaining dry gas to its petrochemical subsidiaries. However, developing a wider infrastructure to bring the gas to a larger number of consumers was not one of Petrobras's priorities, especially as natural gas would compete with the oil products produced by its refineries, in particular the high-sulphur heavy fuel oil, which cannot be exported. Nor was the government, until the mid-1990s, particularly active in promoting natural gas.

**Table 7.6** Natural gas sales by region and state in Brazil, 1992-2001

| mcm | 1992 | 1993 | 1994 | 1995 | 1996 | 1997 | 1998 | 1999 | 2000 | 2001 | 2001 % |
|---|---|---|---|---|---|---|---|---|---|---|---|
| **Northeast** | 1,258 | 1,555 | 1,543 | 1,732 | 1,801 | 1,898 | 2,015 | 2,211 | 2,526 | 2,599 | 29% |
| Ceará | 18 | 21 | 28 | 30 | 35 | 36 | 46 | 61 | 74 | 91 | 1% |
| Rio Grande do Norte | 14 | 13 | 12 | 19 | 26 | 31 | 34 | 38 | 48 | 56 | 1% |
| Paraíba | 5 | 8 | 15 | 16 | 20 | 31 | 34 | 44 | 59 | 69 | 1% |
| Pernambuco | 149 | 131 | 170 | 187 | 182 | 195 | 202 | 212 | 239 | 264 | 3% |
| Alagoas | 124 | 90 | 75 | 108 | 131 | 126 | 147 | 168 | 143 | 145 | 2% |
| Sergipe | 409 | 423 | 412 | 389 | 399 | 436 | 411 | 439 | 512 | 443 | 5% |
| Bahia | 540 | 870 | 831 | 983 | 1,008 | 1,042 | 1,141 | 1,250 | 1,453 | 1,531 | 17% |
| **Southeast** | **1,621** | **1,869** | **2,015** | **2,277** | **2,559** | **2,833** | **2,774** | **3,138** | **3,783** | **4,971** | **56%** |
| Espírito Santo | 130 | 160 | 152 | 163 | 199 | 206 | 221 | 219 | 263 | 337 | 4% |
| Rio de Janeiro | 1,116 | 1,232 | 1,177 | 1,191 | 1,194 | 1,242 | 1,161 | 1,307 | 1,548 | 1,990 | 22% |
| São Paulo | 376 | 477 | 687 | 921 | 1,096 | 1,231 | 1,202 | 1,359 | 1,668 | 2,293 | 26% |
| Minas Gerais | – | – | – | 2 | 70 | 154 | 190 | 253 | 305 | 353 | 4% |
| **South** | **–** | **–** | **–** | **–** | **–** | **–** | **–** | **–** | **262** | **1,239** | **14%** |
| Paraná | – | – | – | – | – | – | – | – | 53 | 127 | 1% |
| Santa Catarina | – | – | – | – | – | – | – | – | 76 | 218 | 2% |
| Rio Gr. do Sul | – | – | – | – | – | – | – | – | 134 | 895 | 10% |
| **Centre-West** | **–** | **–** | **–** | **–** | **–** | **–** | **–** | **–** | **–** | **54** | **1%** |
| Mato Grosso | – | – | – | – | – | – | – | – | – | 54 | 1% |
| **Total Brazil** | **2,879** | **3,424** | **3,559** | **4,010** | **4,360** | **4,731** | **4,789** | **5,349** | **6,572** | **8,864** | **100%** |

*Source: ANP.*

Apart from the supply issue and the lack of support for gas from Petrobras and policy-makers, other structural reasons – mainly the lack of gas demand – prevented earlier and broader expansion of gas use in Brazil:

■ limited potential demand for gas in the residential, commercial and public sectors, as there is no need for space heating in Brazil;

■ in industry, competition with biomass, which is a by-product that would otherwise have to be disposed of as waste, and with low-cost, high-sulphur heavy fuel oil;

■ ample and low-cost hydropower for electricity generation.

**... driven largely by the industrial sector**

Until the end of the 1960s, most of the gas that was not used as fuel on oil platforms was reinjected or flared. Gradually, some natural-gas processing units were installed to recover condensates (butane and propane, used in the production of liquid petroleum gas, as well as natural gasoline). The increasing availability of "dry" gas from those processing units[24] led in the early 1970s to the creation of a first petrochemical complex

24. Dry gas is what is left when the heaviest (and most valuable) fractions (propane, butane, etc) have been extracted from "wet" gas (i.e. gas as it comes out of the field, after dehydration and desulfurisation). The "dry gas" is composed mainly of methane with small percentages of ethane.

in Bahia. Gradually, use of gas by the petrochemical and steel industries increased in the Northeast.

Only at the beginning of the 1980s, following the two oil shocks and the discovery of substantial natural gas reserves in other Northeastern states and in the Southeast, gas consumption spread nationally and into other energy-intensive industries (e.g. glass, ceramics, paper and cellulose, food and beverages, cement and non-ferrous metals). By 1990, the industrial sector was the largest user of natural gas in the country.

In 2000, the industrial sector used 5.2 bcm of gas, accounting for 58% of the country's total primary gas supply. The largest users of natural gas in the industrial sector are the chemical and petrochemical sector (25%), the iron and steel sector (9%), the cement industry (4%) and the pulp and paper industry (3.5%). However, gas only accounts for 6% of the industrial sector's total energy needs. By comparison, biomass accounts for over five times as much, or 33% of total energy use in industry. As the biomass is largely a by-product of food production, and must be disposed of somehow, it seems difficult for gas to compete with biomass. However, the 7.5 Mt of heavy fuel oil and 8.1 Mt of naphtha and gas oil consumed in 2000 by the industrial sector could potentially be replaced by natural gas, driven by environmental considerations in the case of heavy fuel oil (in view of its high sulphur content) and by price considerations in the case of naphtha and gas oil.

**Limited potential for gas use in the residential and commercial sectors**

Because there is virtually no need for space heating in Brazil, gas use in the residential and commercial sectors remains limited to cooking and water heating, making it uneconomic to develop urban gas distribution networks. Demand for gas in the public/commercial and residential sectors started at the beginning of the 1990s and is still concentrated in just two cities: Rio de Janeiro and São Paulo. In both cities, the existence of a distribution network for manufactured gas allowed for relatively rapid growth of gas consumption in the residential and commercial sectors, though demand is now stabilising and new investment in the distribution networks will be necessary to expand the market. In other cities with no pre-existing distribution networks, the cost of access for residential and commercial consumers is much higher.

In 2000, the residential and commercial sectors consumed just 200 mcm of natural gas. New applications, such as use of gas for air-conditioning, refrigeration and distributed power generation could increase urban gas use in the future.

**Automotive gas has a promising future**

The use of gas for transport (compressed natural gas – CNG[25]) reached 313 mcm in 2000. According to preliminary data, this amount nearly doubled in 2001 to about 600 mcm, and is expected to have doubled again in 2002. Most CNG use is concentrated in Rio de Janeiro (40%) and in São Paulo (20%), with substantial development taking place in several states of the Northeast. Currently, some 360 CNG retailing stations exist in Brazil, with over 200 in just two states: Rio de Janeiro and São Paulo. Market growth is encouraged by the competitive price of CNG, which costs only around 60% of the price of regular gasoline.

---

25. In Brazil and other Latin American countries, one finds the expression *vehicular natural gas* abbreviated as VNG.

At the end of 2001, 285,000 vehicles had been converted to CNG in Brazil, or 1% of all vehicles. This number is expected to rise to 465,000 by the end of 2002 and to one million by 2005. By comparison, there are 722,000 CNG vehicles in Argentina, or about 6% of all vehicles. Most CNG vehicles are fleet vehicles, especially taxis. In the state of Rio de Janeiro, taxis are obliged by a state government decree to convert to CNG. The state government now wants to pass a law that would oblige other public-transport vehicles, such as buses and vans, to use natural gas. Currently, converting a gasoline engine to use CNG costs approximately US$790. Taxi drivers in Rio de Janeiro have been asking manufacturers to begin production of CNG vehicles.

Considerable investment in CNG is underway. Gaspetro estimates that investment in CNG supply infrastructure could total US$1.2 billion between 2002 and 2005. The greatest part of this (US$900 million) would be invested in services and equipment for the conversion of gasoline vehicle, while the remaining US$300 million would be needed for in expanding the network of fuel stations. Industry sources project that the number of CNG pumps could triple by 2005, to over 1,000 across the country, increasing total natural gas demand by CNG vehicles to over 3 bcm per year by 2005.

## Gas demand in power generation

With the continued improvement of the economic and thermal efficiency of combined-cycle technology since the mid-1980s, and growing concern about climate change and local air pollution, gas use in power generation has gained momentum worldwide, especially in countries with large gas reserves of their own. In countries and regions wishing to develop their gas markets from scratch, the power-generation sector may serve as an anchor for the development of a gas infrastructure, as the large concentrated demand of a power plant allows substantial economies of scale for pipelines. Long-term gas supply contracts for power stations provide high and predictable demand and a guaranteed stream of income to finance large investments in new transmission pipelines.

In Brazil, however, the abundance of cheap hydropower has hindered the development of thermal generation. The dominance of hydropower created a strong "hydropower culture", which marked a whole generation of engineers and policy makers. The country has a well-developed domestic hydropower industry, capable of managing all parts of a large hydro system. This has led to the emergence of a powerful "hydropower lobby", with considerable economic interests linked to the construction of large dams, which make it difficult for thermal generation to emerge, even where it is economically viable.

Since the mid-1990s, the Brazilian government has been calling for a larger share of thermal capacity, to be fuelled mainly with gas, in order to reduce dependence on hydroelectricity and to boost gas demand. The government expected that new private gas-fired generation would advance the penetration of natural gas in Brazil, bringing the share of gas in total primary energy supply to about 10% by 2010. However, little happened until the 2001 electricity crisis. The government's incentives to promote gas-fired generation and the special challenges of introducing gas-fired power generation in Brazil are dealt with in the section on *Prospects and challenges for gas-fired generation*, later on in this chapter.

## Gas demand projections to 2020

The IEA's projections for natural gas demand in Brazil to 2020 are summarised in *Figure 7.5* and *Table 7.7*. The basic macroeconomic assumptions for Brazil for the period 2000-2020 are a 3% GDP growth and a 1% population growth.

Total primary gas supply is expected to reach 21 Mtoe in 2010 and 36 Mtoe in 2020, compared to 7.5 Mtoe in 2000. This is equivalent to an average annual rate of 8.1% growth over the 20-year period, a much faster growth than total energy supply, which is projected to grow at 2.6% per year over the same period.

The bulk of the increase in gas demand will come from the power-generation sector. Gas-fired generation capacity is expected to increase from around 1 GW in 2000 to 12 GW in 2010 and 24 GW in 2020, generating 56 TWh in 2010 and 104 TWh in 2020, compared to just 2 TWh in 2000. This will increase the share of gas-generated electricity in total generation from 1% currently to 15% in 2020. Gas use in power generation is projected to rise from 640 mcm in 2000 to 12.5 bcm in 2010 and 23.3 bcm in 2020 (34 and 64 mcm/d respectively), rapidly surpassing gas use in industry. These projections are based on projected electricity demand growth of 3.4% between 2000 and 2010 and 3.0% between 2010 and 2020.

By comparison, the Brazilian government, in its latest Ten-year Expansion Plan, projects the need for 46 GW of new capacity between 2001 and 2010, of which some 13 GW are to come from new thermal power stations. This is based on an assumption of 5.3% growth in electricity demand per year.

Use of natural gas in the industrial sector is expected to continue to grow faster than the sector's total energy demand, reflecting the substitution of natural gas for fuel oil. The IEA projects that natural gas demand by the industrial sector will nearly triple between 2000 and 2020, reaching 14 bcm from the current 5 bcm. This is equivalent to a 5.1% average annual growth rate, while industry's total energy demand is expected to increase by 2.2% per year. Reflecting this trend, in 2020, natural gas is expected to provide about 11% of total energy used by the industrial sector, compared with just 6% today.

**Figure 7.5** IEA projections for energy demand in Brazil, 2000-2020

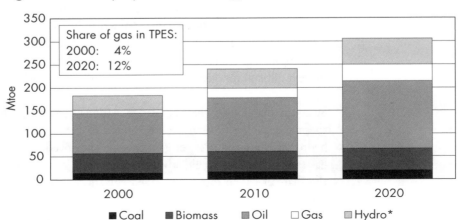

\* *Includes nuclear, solar and wind energy, as well as net imports of electricity.*
*Source: Adapted from IEA, World Energy Outlook 2002.*

Despite projected vigorous growth in demand for gas (4.7% annually over the next 20 years), the IEA projects gas demand by residential and commercial customers to remain relatively marginal, reaching 500 mcm by 2020 (1.1% of total gas supply). As for gas use in transport, IEA projections are much more conservative than those of Brazilian industry: natural gas use in transport is expected to reach about 1 bcm in 2020.

**Table 7.7** IEA projections for natural gas demand in Brazil, 2000-2020

| | Energy Demand (Mtoe) | | | | Growth Rates (% per annum) | | | |
|---|---|---|---|---|---|---|---|---|
| | 1971 | 2000 | 2010 | 2020 | 1971-2000 | 2000-2010 | 2010-2020 | 2000-2020 |
| **Primary Energy** | | | | | | | | |
| **Supply** | **69.6** | **183.2** | **240.8** | **305.0** | **3.4** | **2.8** | **2.4** | **2.6** |
| of which gas | 0.1 | 7.5 | 21.4 | 35.7 | 15.7 | 11.1 | 5.2 | 8.1 |
| % of gas | 0% | 4% | 9% | 12% | | | | |
| **Final Consumption** | **62.7** | **155.8** | **199.2** | **250.5** | **3.2** | **2.5** | **2.3** | **2.4** |
| of which gas | 0.2 | 4.8 | 8.4 | 13.2 | 12.6 | 5.8 | 4.6 | 5.2 |
| % of gas | 0% | 3% | 4% | 5% | | | | |
| **Industry** | **18.1** | **66.9** | **83.1** | **103.8** | **4.6** | **2.2** | **2.3** | **2.2** |
| of which gas | 0.0 | 4.3 | 7.6 | 11.7 | 17.9 | 5.8 | 4.4 | 5.1 |
| % of gas | 0% | 6% | 9% | 11% | | | | |
| **Transportation** | **14.4** | **48.2** | **67.0** | **87.6** | **4.3** | **3.3** | **2.7** | **3.0** |
| of which gas | | 0.3 | 0.5 | 0.9 | | 6.2 | 6.7 | 6.4 |
| % of gas | | 1% | 1% | 1% | | | | |
| **Other Sectors** | **29.0** | **36.2** | **44.0** | **53.2** | **0.8** | **2.0** | **1.9** | **1.9** |
| of which gas | 0.1 | 0.2 | 0.4 | 0.6 | 2.3 | 4.8 | 4.7 | 4.7 |
| % of gas | 0% | 1% | 1% | 1% | | | | |

| | Electricity Generation (TWh) | | | | Growth Rates (% per annum) | | | |
|---|---|---|---|---|---|---|---|---|
| | 1971 | 2000 | 2010 | 2020 | 1971-2000 | 2000-2010 | 2010-2020 | 2000-2020 |
| **Total Generation** | **52** | **349** | **505** | **685** | **6.8** | **3.8** | **3.1** | **3.4** |
| of which gas | | 2 | 56 | 104 | | 37.2 | 6.4 | 20.8 |
| % of gas | | 1% | 11% | 15% | | | | |

| | Power Gen. Capacity (GW) | | | | Growth Rates (% per annum) | | | |
|---|---|---|---|---|---|---|---|---|
| | 1999 | 2010 | 2020 | | – | 1999-2010 | 2010-2020 | 1999-2020 |
| **Total Capacity** | **68** | **102** | **141** | | | **3.8** | **3.3** | **3.7** |
| of which gas | 0 | 12 | 24 | | | 42.5 | 7.0 | 23.5 |
| % of gas | 1% | 12% | 17% | | | | | |

*Source: Adapted from IEA, World Energy Outlook 2002.*

**Natural gas transmission and distribution**

The development of a mature gas market with a diversified range of customers requires a broad pipeline network to bring the fuel to the end-users. As shown in *Figure 7.6*, total gas consumption is closely linked to the length and capacity of the pipeline network.

**Figure 7.6** Evolution of natural gas consumption and network expansion, 1971-2001

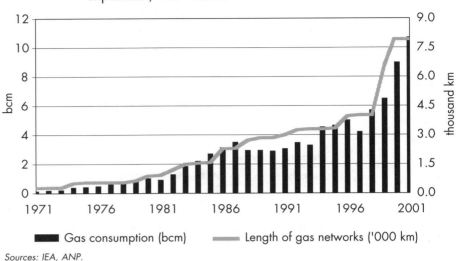

Gas consumption (bcm)          Length of gas networks ('000 km)

*Sources: IEA, ANP.*

Brazil currently has about 8 thousand kilometres of gas pipelines, half of which (including the 2,593-km Brazilian stretch of the Bolivia-Brazil *Gasbol* pipeline) were added in the last three years. As often happens, the Brazilian gas transmission and distribution network started as a number of isolated systems for tapping local gas reserves. Those systems were gradually expanded and interconnected.

There are currently three separate networks in Brazil, two in the Southeast and one along the Northeastern coast. The main Southeastern network supplies the states of Rio de Janeiro, São Paulo and Minas Gerais with gas from the Campos and Santos basins. A smaller network exclusively supplies the state of Espirito Santo from local gas reserves. A pipeline connecting the Espirito Santo network to the main Southeastern network is under study (see *Map 11* at the end of the book).

In addition, the Urucú-Coari pipeline in the state of Amazonas in the Northern region is now ready and likely to be the first segment of a wider network that will supply gas from the Urucú field to several thermoelectric projects in the states of Amazonas and Rondonia. Two extensions of the Urucú-Coari pipeline to supply the cities of Manaus[26] and Porto Velho are currently at an advanced planning stage.[27] Finally, the Bolivia-Brazil *Gasbol* pipeline connects the Southeast network to cities in the Centre-West and the South.

Since the opening of the second part of the Bolivia-Brazil pipeline, stretching from São Paulo to Porto Alegre, all the largest state capitals except Brasilia now have access to natural gas. The main pipelines, their length and capacities are summarised in *Table 7.8* below.

---

26. The Coari-Manaus pipeline may be replaced by river barge transportation.
27. The two cities currently use expensive diesel to generate most of their electricity.

**Table 7.8** Brazil's natural gas transportation system, 2001

| Networks and pipelines | States supplied | Start-up year | Diametre (inches) | Capacity ('000 cm/d) | Length (km) |
|---|---|---|---|---|---|
| **Northeast I** | | | | | |
| Guamaré (RN)-Cabo (PE) | RN, PB, PE | 1986 | 12 | 860 | 424 |
| Guamaré (RN)-Pecém (CE) | CE | 1998-2000 | 12-10 | 800 | 382 |
| Pilar (AL)-Cabo (PE) | AL, PE | 2002 | 12 | 2,000 | 204 |
| **Northeast II** | | | | | |
| Atalaia (SE)-Catu (BA) | SE, BA | 1974 | 14 | 1,103 | 232 |
| Santiago (BA)-Camaçari I (BA) | BA | 1975 | 14 | 1,000 | 32 |
| Santiago (BA)-Camaçari II (BA) | BA | 1992 | 18 | 1,800 | 32 |
| Candeias (BA)-Camaçari (BA) | BA | 1981 | 12 | 1,000 | 37 |
| Aratu (BA)-Camaçari (BA) | BA | 1970 | 10 | 700 | 20 |
| *Pilar (AL)-Atalaia (SE)* | | *Under contruction* | | | |
| **Southeast I** | | | | | |
| Cabiúnas (RJ)-Reduc (RJ) | RJ | 1982 | 16 | 4,250 | 183 |
| Reduc (RJ)-Regap (MG) | MG | 1996 | 16 | 1,952 | 357 |
| Reduc (RJ)-Esvol (RJ) | RJ | 1986 | 18 | 4,215 | 95 |
| Esvol (RJ)-Tevol (RJ) | RJ | 1986 | 14 | 4,215 | 5.5 |
| Esvol (RJ)-São Paulo (SP) | RJ, SP | 1988 | 22 | 4,215 | 326 |
| RBPC (SP)-Capuava (SP) | SP | 1993 | 12 | 1,530 | 37 |
| RBPC (SP)-Comgás (SP) | SP | 1993 | 12 | 1,550 | 1.5 |
| **Southeast II** | | | | | |
| Lagoa Parda (SE)-Aracruz (SE) | ES | 1983 | 8 | 1,000 | 38 |
| Aracruz (SE)-Vitória (SE) | ES | 1984 | 8 | 1,000 | 74 |
| Serra (SE)-Viana (SE) | ES | 1997 | 8 | 660 | 30 |
| *Cabiúnas (RJ)-Vitória (SE)* | | *Under study* | | | |
| **North** | | | | | |
| Urucú (AM)-Coari (AM) | AM | Completed | | | 280 |
| *Coari (AM)-Manuas (AM)* | AM | *Under study* | | | |
| *Urucú (AM)-Porto Velho (RO)* | RO | *Under study* | | | |
| **Cross-border pipelines** | | | | | |
| Gasbol 1 | | 1999 | | | 1,418 |
| Gasbol 2 | | 2000 | | | 1,180 |
| Lateral Cuiabá | | 2001 | | | 267 |

*Sources: ANP, GásEnergia website.*

## Gas transportation companies

*Transpetro*, a wholly-owned Petrobras subsidiary, is responsible for operating the whole transportation network for domestic gas. Transpetro was created in 1998 to comply with the 1997 Petroleum Law, which requires the separation of gas production and transportation facilities into different legal entities. Three other transportation companies operate the Brazilian segments of the existing gas import pipelines: *Transportadora Brasileira Gasoducto Bolivia Brasil S.A.* (TBG), *Gasocidente do Mato Grosso Ltda.* and *Transportadora Sulbrasileira* (TSB). These are described in more detail in the section on *Gas trade*. New pipelines can be built by any company, public or private, and it is likely that many of the new projects will be built by Petrobras in partnership with private companies.

**Gas distribution companies**

Until 1988, the distribution of natural gas was a municipal service, and there were only two natural gas distribution companies: *Companhia de Gás de São Paulo* (Comgas) in São Paulo and *Companhia Estadual de Gás* (CEG) in Rio de Janeiro. Under the new 1988 Federal Constitution, distribution of natural gas became the responsibility of the states. A constitutional amendment in 1995 allowed both private and public companies to obtain licences for the distribution of natural gas.

There are currently 22 distribution companies operating in 18 states. The states of São Paulo and Rio de Janeiro together have five distribution companies (*Comgas, Gas Brasiliano* and *Gas Natural* in São Paulo; *CEG* and *CEG Rio* in Rio de Janeiro). These five distribution companies are 100% privately owned. The other 17 companies have mixed ownership. Many of them follow a tripartite model with the state government owning 51% of the shares, *Gaspetro* 24.5% and private companies the remaining 24.5%. Private companies with stakes in the Brazilian gas distribution sector include *Gas Participações Ltda.* (Gaspart), a subsidiary of *Enron*; UK-based *BG* and *Shell*; Spanish *Gas Natural Sdg.* and *Iberdrola*; Italian *Italgas* and *Snam*; Argentine *Pluspetrol* and several Brazilian industrial consortia.

At the moment, only Comgas, CEG, CEG Rio and Bahiagas have significant distribution infrastructure in place. The other companies are just starting to build up a network or are still awaiting gas supply.

## PROSPECTS AND CHALLENGES FOR GAS-FIRED POWER GENERATION

Brazil's high reliance on hydropower is comparable only to Norway's. Two major differences, however, are that Norway is interconnected with the thermoelectric systems of its neighbouring countries and that the growth rate of electricity consumption is low in Norway, while it will remain high in Brazil for a long time to come, requiring significant additional investment in power capacity in the medium and long term. An important structural question is how new gas-fired power plants would operate in Brazil's hydropower-dominated system. This complex issue is being widely discussed in Brazil. While a full discussion of the problem and possible solutions is beyond the scope of this report, some main aspects are considered below.

**The complexities of introducing gas-fired power in a hydro-dominated system**

When the financial viability of the Bolivia-Brazil pipeline was assessed, it was expected that, like in other countries, new gas-fired power generation would provide high and predictable demand to justify the investment in the early years, while other markets were gradually developed. Certainly, Brazil badly needed new generation capacity, but the market potential of gas-fuelled thermal generation was assessed without duly considering the specificities of Brazil's hydro-dominated power system.

In hydro-dominated power systems, some thermal generation is generally useful as a back-up for periods of low rainfall. Those thermal power plants may however be idle in periods or years of high precipitation. When there is abundant water in the reservoirs, hydropower plants are dispatched before thermal plants because of their lower variable costs and to avoid spilling of water. Under these circumstances, gas-fired gas turbines may be economically attractive and financially viable as back-up plants as they have very low capital costs, and thereby a limited risk, as long as they have no take-or-pay obligation for the gas.

In highly-developed gas markets, it is usually possible for power stations to obtain gas contracts for shorter periods or with more flexible take-or-pay clauses, because of the existence of gas hubs or, at least, of other consumers (usually industry) who are able to switch fuels and absorb the excess gas in periods of low demand from the power sector. This might become possible in Brazil in the future. Today, however, the industrial market for gas is too small to be able to accommodate large variations in gas supply.

In emerging gas markets, the need to recover the cost of large pipeline investments means that gas is generally sold under long-term supply contracts with relatively strict take-or-pay clauses. Hence, project financing for a gas-fired power plant is generally dependent on the signing of a back-to-back, long-term power purchase agreement (PPA) with a distribution company or other electricity user. The PPA is needed to secure the financing not only of the power plant, but also of the transportation infrastructure needed to bring the gas to the power station. One way for the government to ensure that thermal power projects are bankable is to guarantee them a buyer of last resort for their power.

In the absence of a firm PPA, the cost per kWh to be recovered increases in inverse proportion to the load factor. If this higher cost can be passed on to the electricity consumer via higher electricity prices, it should still be possible to finance the project. However, such high prices will generally only be attainable in situations of power shortage. In addition, Brazil's long tradition of maintaining electricity tariffs artificially low, for social and political reasons, may make pass-through mechanisms difficult to enforce.

A solution advocated by some[28] is to compel all suppliers to pay a share of the extra costs of gas-fired thermal generation (as is already done for the power generated by the Paraguayan part of Itaipu, which is paid in US dollars). Each supplier would be obliged to purchase a share of the (higher-cost) gas-fired generation as a condition to be allowed to operate in the market. The measure would be neutral on competition and the extra cost would be passed to the final consumer as a sort "supply security tax". Final consumers' tariffs would therefore increase, but less and with much less variance than in a system where tariffs reflect marginal costs. In any case, the additional insurance cost is likely to be clearly lower than the costs of rationing and emergency power incurred in 2001.

**Volume flexibility in gas take-or-pay contracts is crucial**

Another issue to be addressed is how gas sellers and buyers would share the price and volume risks which would derive from the use of gas in back-up thermal generation. The gas sales contracts covering the import of gas to Brazil are not in the public domain and solutions must be found by the gas contract partners themselves. Under a strict competition approach, the value of the gas in years of abundant precipitation would be zero, whereas it might be very high in years of low precipitation. With a strict

---

28. See, for example, Dos Santos et al. (2000).

minimum-pay provision,[29] this might lead to situations where water is spilled over the dam to fulfil the minimum pay conditions.

An alternative approach would be to determine the value of the gas based on the costs of using gas oil in the newly built CCGTs or fuel oil in other thermal-power plant. In this case, the contractual provisions should allow for enough volume flexibility from one year to the next to adapt the gas off-take to variations in precipitation and in growth of power demand.

The first approach is in line with the "classical" approach, according to which the gas seller takes the price risk and the buyer takes the volume risk. However, even in Continental European gas contracts, it is not unusual for the seller also to bear some of the volume risk, e.g. risks linked to independent factors, such as the weather (temperature), which cannot be influenced by the buyer.

## Special pricing problems of the Brazilian gas import contracts

One of the main problems weighing heavily on gas-fired power projects in Brazil relates to the price of gas. Most of the gas to be used by new thermal power plants is gas imported from Bolivia. Its price is set in US dollars and indexed to a basket of international fuel oil prices. As a result of the sharp increase in international oil prices and the gradual devaluation of the Brazilian currency, the price of Bolivian gas in *reais* has risen substantially since 1999. Currently, city-gate gas prices in São Paulo and other Brazilian cities along the Gasbol pipeline are around US$3.2/million British thermal units (Mbtu), well above prices for Brazilian gas and prices for fuel oil or coal. Prices are also higher than the price Argentine producers are likely to charge, once their gas flows into southern Brazil through pipelines currently under construction or planned. Under these circumstances, power distribution companies and other potential electricity purchasers have been unwilling to sign long-term PPAs with power plant developers.

Another problem is the mismatch between quarterly adjustments of the gas price and annual adjustment of the electricity tariffs. The fact that gas supply contracts have prices fixed in US dollars, while electricity sales are in Brazilian *reais* means that power plant operators take on the risk of a currency devaluation between two annual electricity tariffs adjustments. This has been an obstacle to obtaining financing, as banks are concerned that in the case of a major devaluation of the Brazilian currency and/or a large increase in international oil prices, power project operators might default on their loans.

The absence of competition in gas supply is another concern. Despite the liberalisation of the upstream oil and gas sector, Petrobras is currently the sole producer of domestic gas and the dominant importer of gas. It sells virtually all Bolivian gas due to its preferential access to the Bolivia-Brazil pipeline, an agreement which predates – and has conflicted with – the new regulations on third party access (TPA) to pipelines. Power developers have been unwilling to lock themselves into 20- to 30-year take-or-pay gas supply contracts with Petrobras, believing that additional pipelines from

---

29. A minimum-pay provision is a clause in the take-or-pay contracts that obliges the buyer to take a minimum amount of gas.

Argentina, and possibly from Bolivia, would bring prices down over time. Indeed, BG, which has been allowed access to Bolivia-Brazil pipeline as long as there is spare capacity, has been able to the undercut Petrobras's city-gate price, by offering a lower commodity price coupled with a lower transportation charge.

## The Thermoelectric Priority Program

In September 1999, the government – concerned about the lack of investments and the accelerating demand for electricity due to economic recovering – announced a series of measures to solve some of the difficulties faced by private developers in obtaining financing for gas-fired power projects. The incentives included: a) guaranteed gas supplies with an average price ceiling of US$2.26/MBtu for 20-year supply contracts;[30] b) the availability of long-term PPAs from state-owned Eletrobras, which will be the ultimate buyer of the electricity generated; c) the possibility of obtaining special favourable loans from the state-owned national development bank, BNDES, for the acquisition of domestically-produced equipment.[31] The government expected that these measures would promote the construction of some 23 thermal power plants (all but three gas-fired), with a total capacity of 7.5 GW.

In February 2000, the above measures were translated into law under the name *Programa Prioritário de Termeletricidade* (Thermoelectric Priority Programme – PPT).[32] Eligible projects (see *Box 7.3*) are required to be operational by the end of 2003. In July 2000, PPT projects were offered the possibility of paying a slightly higher gas price (US$2.475/MBtu) in exchange for an annual gas price readjustment in line with that of electricity tariffs.

Only a few weeks after the PTT was launched, additional incentives were announced for plants able to start operations by December 2001, including tax breaks. This fast-track programme, named Emergency Thermoelectric Programme, included 10-12 plants, with state-owned Petrobras called to finance most of the projects.

Still, by mid-2001, when electricity rationing became inevitable, not a single PPT gas-fired power plant was ready, and work had started on just five. In June 2001, the government finally agreed to take one last step, assuming – through Petrobras – the foreign exchange risk between two annual gas price adjustments. The new measure mandates that Petrobras convert the US dollar gas price into *reais* at the beginning of each year, and re-evaluate it only at the end of the year. The difference between the average price in US dollar throughout the year and the value set in *reais* at the beginning of the year accumulates in a "compensation account". It is then added (or subtracted, in the case of a re-valuation of the *real*) to the new value fixed in *reais* for the following year. The rule applies to the first 40 mcm/d contracted with thermal plants operating before 2003.

---

30. In the Northeast, where gas will at least initially be sourced domestically, the ceiling will be lower.
31. These incentives effectively give IPP projects the character of BOO projects.
32. Ministerial Decree 3.371 of 24 February 2000, Regulation MME No. 43 of 25 February 2000.

**Box 7.3**        The Thermoelectric Priority Programme

The PPT originally included 53 power projects, of which 47 are gas-fired. These latter consisted of 30 combined-cycle gas turbines (CCGT), 2 simple cycle gas turbines (GT), 11 cogeneration plants and 4 conversions of existing oil plants to natural gas, with a total capacity of 17.6 GW. Other projects were added to the PPT list later, and some dropped out. The original deadline for the completion of the plants was extended from December 2003 to December 2004.

As of April 2002, the PPT included 40 gas-fired CCGT power stations totalling 13.6 GW, of which 6 were already in operation, 4 were undergoing pre-operation tests, 11 were under construction, 8 were about to start construction, 9 had not yet started construction and might be dropped from the programme, and 2 were new projects. If all projects were completed on schedule, by the end of 2004 they would be expected to consume some 58 mcm/d (21 bcm per year) of natural gas (see *Table 7.9*).

Total investment is estimated at US$9.6 billion, of which 76% will come from private developers, 20% from Petrobras and 4% from the state-owned electricity company Eletrobras. Eighteen plants (7.5 GW) are 100% private, while 17 others (5 GW) have mixed ownership, with the private partner usually holding a controlling stake. Petrobras participates in 15 private power projects (4.7 GW), with controlling stakes in two of them, and is the sole investor in two other projects (400 MW). Eletrobras, through its subsidiaries CHESF and Furnas, is developing two power projects alone (800 MW) and has small shares in two private projects (273 MW).

Of the six PPT plants already in operation, two were built to supply the spot market, but all the others have signed PPAs for at least a portion of their output. The total amount of power to be supplied on the spot market is 1.4 GW. Petrobras has agreed to buy 2.8 GW from the new plants under long-term PPAs. Of this amount, the company's internal consumption is 276 MW, and some 1,000 MW have already been sold to customers. The remaining 1,550 MW are under negotiation.

Most of the PPT projects will supply power to the interconnected system comprising the South, Southeast and Centre-West regions: 27 projects with a combined capacity of 10.7 GW. Twelve plants with a capacity of 2.5 GW will supply the Northeast, while only one project of 400 MW is located in the North.

**The situation after the power crisis in 2001**

The lack of investment in new power-generation capacity at a time of rapid demand growth and the fact that new gas-fired capacity was not yet operational during the draught in 2001 resulted in a severe power shortage. The consequent demand-side management measures – i.e. raising tariffs – and the resulting (lasting) drop in electricity demand, combined with above-average rainfall in 2002 caused overcapacity of electricity supply. This in turn led to an oversupply of gas, especially with a gradual increase in domestic gas production and the increasing take-or-pay obligations for the gas from Bolivia.

**Table 7.9** Current status of the Thermoelectric Priority Programme, as of April 2002

| Status as of April 2002 | N° | Generating capacity at end year (MW) | | | | Investment (US$ billion) | | |
|---|---|---|---|---|---|---|---|---|
| | | 2001 | 2002 | 2003 | 2004 | Total | Priv. | Public |
| (a) In operation | 6 | 1,353 | 1,721 | 1,901 | 1,901 | 1.1 | 1.1 | 0.0 |
| (b) Pre-operation testing | 4 | 0 | 1,000 | 1,000 | 1,070 | 1.3 | 0.5 | 0.8 |
| (c) Under construction | 11 | 0 | 1,336 | 3,783 | 4,279 | 2.9 | 2.0 | 0.9 |
| (d) Construction about to start | 8 | 0 | 215 | 1,930 | 2,280 | 1.5 | 1.2 | 0.3 |
| **Total (a) + (b) + (c) + (d)** | **29** | **1,353** | **4,272** | **8,614** | **9,530** | **6.8** | **4.8** | **2.0** |
| Gas use (bcm/year) | | 3.1 | 7.5 | 14.0 | 14.8 | | | |
| (e) Construction not yet started | 9 | 0 | 0 | 150 | 2,642 | 1.8 | 1.5 | 0.3 |
| (f) New projects to be included | 2 | 0 | 0 | 330 | 1,465 | 1.0 | 1.0 | 0.0 |
| **Total (e) + (f)** | **11** | **0** | **0** | **480** | **4,107** | **2.8** | **2.5** | **0.3** |
| Gas use (bcm/year) | | 0.0 | 0.0 | 1.0 | 6.3 | | | |
| **Grand total** | **40** | **1,353** | **4,272** | **9,094** | **13,637** | **9.6** | **7.2** | **2.3** |
| Gas use (bcm/year) | | 3.1 | 7.5 | 15.1 | 21.1 | | | |

Source: MME.

The challenge is to bring the following three variables into balance: (i) the average hydropower generation capacity, which depends on precipitation; (ii) electricity demand; and (iii) the production from a thermal system based mainly on imported gas.

(i) The draught in 2001 showed how much the capacity of the hydro system can vary. Its average capacity was about 10 GW less in 2001 than in a year with average precipitation.

(ii) Electricity demand, which was characterised by a long period of strong growth, reacted more than expected to the tariff increases after the power crisis. While demand picked up again in 2002 and almost reached the pre-crisis level, there remain some question marks concerning the future rate of growth in power demand and possible demand reactions to tariff increases.

(iii) The thermal-power system seems ideal for providing a balance between a rapidly-growing power demand and a power supply based on hydro. However, the capacity to be provided by the thermal system depends on the gap between power demand and power supply by the hydro system, which may vary greatly. The PPT foresees the construction of 13.6 GW of gas-fired power capacity by 2004. By the end of 2001, however, less than 2 GW had come on stream. How much of the remaining gas-fired capacity will be completed, and when, depends on private investors. The gas supply for the power stations requires sufficient investment in gas production and transportation capacity to meet the stations' maximum demand. The capacity of the Gasbol pipeline, which will soon reach 30 mcm/d, is certainly more than sufficient to supply gas to the gas-fired power plants, which are already or will soon become operational. However, its capacity would probably not be sufficient to supply gas for all the gas-fired power plants foreseen under the PPT. Additional gas pipelines would have to be built.

The challenge is to provide the right incentives for timely investment to ensure a secure supply of electricity while avoiding large over-investment in any part of the gas or power system.

## GAS-SECTOR REFORM

Until 1995, Petrobras had a legal monopoly on all activities along the natural gas chain, with the exception of gas distribution, which was the responsibility of state-level public companies. In 1995, in the context of wider structural reforms aimed at accelerating the opening of the Brazilian economy to private investment and international competition, the Brazilian Congress approved two amendments to the federal constitution effectively removing the restrictions on private activity in the hydrocarbons sector. Amendment No.9 allows private companies to participate in all *upstream* (exploration, development and production) and *midstream* (imports and exports, processing, refining, storage and transportation) oil and gas activities. Amendment No.5 allows private companies to carry out the distribution and commercialisation of natural gas.

In 1997, a new Petroleum Law (Law 9478) was enacted after nearly two years of contentious political debate. The Petroleum Law sets out the new "rules of the game" and creates the *Agência Nacional do Petróleo* (ANP). The ANP, an autonomous body of the Federal Public Administration linked to the Ministry of Mines and Energy (MME), was given broad responsibilities for overseeing the transition from a vertically-integrated, state-controlled sector to a competitive sector able to attract private investment. Like other regulatory agencies, the ANP is in charge of fostering competition in the competitive segments of the oil and gas industry, supervising and regulating the non-competitive segments and protecting consumers. However, the ANP's tasks also include the granting of exploration and production concessions through competitive bidding rounds – a role not normally performed by regulatory agencies.

There are five important points to note about the reform of the oil and gas sector in Brazil:

■ The Federal government remains the sole owner of all oil and gas reserves in the Brazilian subsoil,[33] and retains the rights over their exploitation, which it grants – through concessions or authorisations – to public and private companies, domestic or foreign. Hence, Petrobras no longer enjoys exclusive rights and must now compete on equal

---

33. Everywhere in the world, natural resources are vested in the state, except for the US where natural resources belong to the owner of the land above the resources.

footing with all the other players in the market. Similarly, Amendment No.5 does not remove the states' monopoly over gas distribution activities, but allows them to award concessions to public, private or mixed-capital companies.

■ Unlike the electricity reform, there is no question of downsizing and privatising the state oil company Petrobras, as was done for example in nearby Argentina. The 1997 Petroleum Law makes it very clear that control of Petrobras is to remain in public hands. Article 62 stipulates that the federal government must retain a majority of Petrobras's voting shares. However, Petrobras is allowed to enter into joint ventures with private companies and to create subsidiaries, in which it may be a minority shareholder.

■ The new Petroleum Law does not introduce restrictions on the vertical integration and cross-ownership of companies involved in the various activities in the gas chain. There is a requirement to separate production and transportation activities into different legal entities, but no restriction about cross-ownership of these entities.

■ The Petroleum Law grants third party access to all new and existing pipelines, port facilities and terminals. The conditions of access can be freely negotiated between market players.

■ ANP's responsibilities with regard to natural gas extend from exploration and production, imports and exports, to transportation up to city-gates. From there on, i.e. for gas distribution and commercialisation, regulatory authority lies with the state governments, most of which have state regulatory agencies covering various public services (electricity and gas, but also telecommunications, transport, sanitation, water, etc.). This creates difficulties in establishing a coherent regulatory framework for the entire gas sector, as reform in this segment is likely to evolve differently in each of the 27 states.

**Production**

The first step towards increased competition in the upstream oil and gas sector, was the ratification – through a concession contract with the ANP – of Petrobras' existing E&P activities. The 1997 Petroleum Law allowed Petrobras to retain all its production assets and the right to continue to explore and develop those areas where significant investment had already taken place. In mid-1998, Petrobras signed concession agreements with the ANP for 397 areas: 231 producing areas, 51 areas under development and 115 areas under exploration.[34] In all, they account for 100% of existing oil and gas production and over 7% of the country's total acreage.

The rights to explore the other blocks not requested by or conceded to Petrobras, including those relinquished by the company for lack of exploratory results,[35] are to be offered by the ANP to the highest bidder in annual auctions. So far there have been

---

34. The concession contracts signed by Petrobras are for a production period of 27 years that can be extended at the ANP's discretion.
35. For the blocks not yet under production, Petrobras was allowed three years to reach "commerciallity". After that period the concession would be terminated and the blocks would revert to the ANP. As a result of this provision, 58 exploration blocks granted to Petrobras in 1998 reverted to the ANP in August 2001.

four bidding rounds, resulting in 49 companies' (including Petrobras) signing new exploration concessions for 88 blocks. Private companies bid alone or in consortia, often including Petrobras. The ANP expects that these concessions will generate US$4 billion in new investment in the next ten years.

Despite the fact that participation in the upstream sector is now much more diversified, it will take time for private companies to explore and develop their blocks. Hence, Petrobras will remain the dominant oil and gas producer for some time.

## Imports

While new domestic gas producers will emerge in the longer term, in the short and medium term the only way to increase competition in gas supply is through imports. The 1997 Petroleum Law stipulates that any company or consortium, private or public, can import (or export) natural gas provided it is authorised by the ANP. However, importing gas requires the use of existing pipelines or the construction of new ones. Hence, regulation of gas transportation is the most crucial element in Brazil's natural gas reform.

## Transportation

The construction, extension, and operation of gas transportation pipelines can be carried out by any company, or consortium registered in Brazil and authorised by the ANP.

As mentioned previously, production and transportation facilities must be unbundled into separate legal entities, but there are no restrictions on cross-ownership of these entities. Hence, in 1998 Petrobras transferred all its oil and gas transportation assets to Transpetro, a wholly-owned subsidiary. Transportation companies are restricted to selling transportation services. They cannot buy or sell gas.

Under the open-access regime, the transporter must communicate to the ANP and to the market its "available capacity" (maximum capacity minus gas supply contracts) as well as its "idle contracted capacity" (contracted capacity minus volume of gas transported daily). If there is available capacity, the transporter is obliged to transport gas belonging to other companies, on a firm or interruptible basis, at "market prices". Only interruptible service may be offered by the transporter on the basis of idle contracted capacity. The conditions of access, in particular the tariff to be paid by the user (shipper) to the owner of the facility (transporter), are to be negotiated between the parties, with the ANP mediating or intervening only in case of conflict.

## Distribution

The downstream segment of the gas industry is under the jurisdiction of state governments. Hence, regulation for distribution and commercialisation of gas has evolved differently in different states. In some states, for example, large gas consumers are allowed to buy their gas directly from the producers. In this case, the user can either physically bypass the distribution network by building his own supply pipeline from the transportation pipeline, or he can commercially bypass the distribution company by contracting independently for the gas and transportation and paying the distribution company for the use of its network.

## Prices and tariffs

Domestically-produced gas is treated differently from imported gas. Until 1999, the bundled price of domestic gas delivered at the city-gates (i.e. the price paid by the local distribution companies) was subject to a ceiling linked to the price of fuel oil. In

February 2000, a new tariff policy was introduced which separated the price of gas into two clearly different components: the price of the commodity and the transport tariff. ANP establishes "reference transport tariffs" on a cost-plus basis, varying according to distance and volume transported. These "reference transport tariffs", which are readjusted annually, are used to calculate the ceiling for the prices paid by distribution companies. The commodity price is regulated by the MME in cooperation with the Ministry of Finance and is readjusted quarterly.

The price of imported gas is not regulated although, as mentioned earlier, the government has established special pricing rules for gas used by the power stations included in the Thermoelectric Priority Programme. The transport tariff for the Gasbol pipeline is determined on a postage-stamp basis, i.e. it does not vary with distance, and it is subdivided into a capacity charge and a volume charge.

## CHALLENGES AND UNCERTAINTIES

**Need for a comprehensive energy policy**

In the last few years, Brazil has suffered from the *lack of a comprehensive energy policy*, addressing simultaneously aspects related to the electricity, gas and oil sectors, and from the temporary need for crisis management. Insufficient resources and technical expertise, and a high turn-over in the senior and mid-level positions have reduced the ability of the MME to play an effective role in defining a coherent energy policy and in monitoring energy developments in order to react quickly to market failures and other crises.

**Issues to be addressed by a comprehensive energy policy**

The government should clearly define the *roles of the players* and provide policy-makers with sufficient staff. The regulatory agencies should concentrate on policy implementation and should not have to fill in gaps in policy design. The work of the regulatory agencies dealing with gas transportation, gas distribution and electricity should be co-ordinated.

Another issue to be addressed is the *sequencing of market reforms* and further steps in privatisation. A clear policy on further market reforms is necessary before proceeding to further privatisation. Petrobras has now been partially privatised; however, it still has a dominant position in gas production, import, transportation and distribution in Brazil. The regulatory agencies should be strengthened to enable them to implement the existing rules in any dispute with a strong incumbent, such as Petrobras.

The next steps of market reform should take into account the *specific features of Brazil's energy sector*, especially the hydro-dominated electricity sector and how it affects the introduction of gas for power generation. While it is necessary to take stock of previous decisions, prior commitments by the public sector to private investors should be honoured, to maintain and improve Brazil's credit standing with the international financing community. This does not mean that the government should bail out investors who deliberately took risks.

A comprehensive energy policy should also address other issues, such as environmental objectives, fiscal policy linked to energy taxation, and social and employment questions.

## Dealing with the fall-out of the power crisis

Now that the power crisis is over, there is a need to take stock of the new situation, characterised for some years by overcapacity in gas-fired power generation and, correspondingly, in gas infrastructure. Given a potential slowing of growth in power demand, it may take some time until the market absorbs that overcapacity.

While the adaptation of gas import contracts to the new situation has to be left to the private partners involved, the issue of how to respect the interests of Bolivia, mainly with regard to its upstream income, may have to be addressed by the Brazilian government.

## Co-ordinating further development of the power and the gas sectors

In view of the continuing increase in power demand in the medium and long term, emphasis should be placed on creating the appropriate incentives for investments in new power-generating capacity, to ensure long-term security of supply despite variations in hydrology (for example, customers might be charged an insurance premium for the extra (thermal) capacity needed). However, the size of a gas-fired power system in relation to the hydro system, and the implications for the gas infrastructure, should be reassessed.

The sharp decrease in demand provoked by the tariff increases in 2001 showed the electricity-saving potential on the demand side. The electricity tariffs should provide incentives for demand-side reaction and should ensure that the costs of investment in new power plants can be recovered from tariffs.

Consideration should be given to what steps need to be taken before Brazil can create a wholesale electricity market and how to secure investment in a thermal-power system. Brazil might usefully observe further developments in systems like Norway's. An important issue to be addressed is how to deal with old hydro plants whose book value is very low.

With regard to gas regulation, the existing negotiated TPA regime should be implemented more effectively. This requires rules for a reply to TPA requests within a reasonable time frame of a few weeks and the ability of the regulator to deal with disputes, also within a few weeks, in order to allow for proper competition.

# CHAPTER 8
# VENEZUELA

Venezuela, the world's sixth-largest oil producer, is moving to capitalise on its enormous natural gas resources. The country holds four trillion cubic metres of proven gas reserves, the eighth-largest gas reserves in the world and the largest in South America. However, 91% of Venezuela's proven gas reserves are associated with oil, and much of the gas produced is reinjected into oil wells to boost crude oil production. Hence, gas production is very dependent on OPEC production quotas. The government is now pushing to develop non-associated gas reserves and increase the share of gas in the energy balance. The 1999 Gas Law opens the door to private investment in all gas-related activities except exploration and production of associated gas. The first-ever Venezuelan gas-only E&P licenses were granted in June 2001 to private consortia which include four foreign and two local companies. Despite the undoubtedly large potential for natural gas development, investors' response has been subdued. Concerns remain over the issues of gas pricing, the government's focus on the development of local markets rather than exports, and – last but not least – the country's current political and economic uncertainties.

## Venezuela at a glance, 2000

| | Venezuela | Share in region |
|---|---|---|
| Surface area: | 912,050 km² | 5% |
| Population: | 24 million | 7% |
| Capital: | Caracas | |
| Currency: | Bolivar (Bs) | |
| GDP*: | US$133 billion | 6% |
| Total primary energy production: | 225 Mtoe | 38% |
| Primary energy consumption: | 59 Mtoe | 15% |
| Primary gas consumption: | 27 Mtoe | 31% |
| | **Venezuela** | **Regional average** |
| Per capita GDP*: | US$5,518 | US$6,821 |
| Per capita primary energy cons.: | 2.45 toe | 1.15 toe |
| Per capita electricity consumption: | 2,669 kWhw | 1,708 kWh |

\* *Gross domestic product (GDP) expressed in 1995 prices and PPP.*

## BACKGROUND

Venezuela lies in the far north of South America. It is the sixth-largest South American country and the fifth most populous, with 7% of the continent's total population. It is bordered to the north by the Caribbean Sea and the Atlantic Ocean, on the west by Colombia, on the south by Brazil and on the east by Guyana. With a population of 24 million, Venezuela is not very densely populated (27 people per square kilometre). Population is concentrated on the coast and in the main urban centres. Urbanisation is high, at 87%, compared with 79% on average for South America.

Venezuela has enormous natural resources, but ill-conceived economic policies over the past two decades have led to extremely poor economic performance. Successive governments have lacked the political and ideological commitment to implement structural reforms. The Venezuelan economy remains highly dependent on oil, its mainstay since the 1920s. In 2001, oil accounted for 78% of total export earnings, about half of government revenues and 28% of GDP. This dependence has led to a boom-bust economic pattern, as governments have pursued expansionary policies when oil prices were high, then encountered fiscal and balance-of-payments difficulties when oil prices weakened. The 1998 slump in international oil prices dealt a severe blow to the economy, and the price recovery of 1999 was insufficient to save Venezuela from a 6% contraction in GDP.

On 6 December 1998, Hugo Chávez Frías, a former army lieutenant and author of the 1992 failed military coup, won the presidential elections with over 56% of the vote. His victory ended 40 years of political dominance by the established parties Democratic Action (AD) and Copei. Chávez won overwhelming popular support, especially – but not only – from Venezuela's poor, pledging to end corruption and mismanagement, revitalise the economy and increase standards of living through a "peaceful revolution".

Although he inherited a critical economic situation, President Chávez gave priority to political and institutional reform over the development of a coherent economic agenda. Within six months of taking office in February 1999, Chávez obtained from Congress special enabling powers to legislate by decree, dissolved the Congress and created a 131-member Constituent Assembly. The Assembly's main task was to rewrite the country's 1961 constitution. The new constitution, which was approved by 71% of the voters by referendum in December 1999, increased the presidential term from five years to six and allowed the president to run for re-election. It created the new post of vice-president (to be designated by the president) and replaced the bicameral Congress of the Republic with a unicameral National Assembly.

Chávez's popularity has been decreasing starkly over the last three years. There is growing frustration among his supporters over the lack of progress in reducing crime, creating jobs and curbing corruption. Continuing political conflict has taken a toll on policy-making, and little progress has been made on the economic agenda. In November 2000, Chávez once again obtained, through an "Enabling Law", special powers to legislate by decree for 12 months. This allowed him to pass a number of laws by November 2001 without congressional approval, including the controversial Hydrocarbons Law. In February 2002, Chávez's opponents staged a *coup d'état* that failed. In December 2002, the country was crippled by a ten-day strike in opposition to Chávez.

Continuing high oil prices helped revive growth in 2000 (+3.5%), but the country's economic outlook for 2001-2002 is highly dependent on high oil prices to support the expansion of public expenditure. Despite Chávez's pledge to make his country more attractive to private investors,[1] private investment is unlikely to play a major role in recovery. Uncertainty about government policy and the overvaluation of the national

---

1. For example, in 1999, he issued a Law for Promotion and Protection of Investment, which guarantees stability in taxation and investment incentives for up to ten years after a contract is signed.

currency, coupled with Chávez's fierce nationalistic rhetoric and increasing concentration of power, have eroded investors' confidence.

**Venezuela's new role in OPEC**

Venezuela is a founding member of the Organization of Petroleum Exporting Countries (OPEC) and one of the world's largest oil exporters. While Venezuela's adherence to OPEC production quotas had been erratic in the past, the Chávez government has kept the country's production very close to agreed quota levels. Venezuela's attitude has contributed greatly to OPEC's new-found cohesion and to the success of its strategy of cutting production.

**Regional integration**

Apart from reinforcing links with OPEC partners, Chávez's foreign policy has been geared toward forging new ties and extending trade relationships with neighbouring countries in an effort to reduce Venezuela's commercial dependence on the United States.

Venezuela is a member of the Andean Community (CAN), which includes Bolivia, Colombia, Ecuador and Peru. While Venezuela opposes US plans to move up the creation of a Free-Trade Agreement of the Americas (FTAA) from 2005 to 2003, the government is keen to promote stronger economic integration with South America. In 2001, Venezuela submitted its application to become an associate member of the Mercado Común del Sur (Mercosur), a block whose main industrial and population centres lie several thousand kilometres south of Venezuela's borders. This application, which is supported by Brazil, has caused controversy within the CAN, which is currently negotiating multilateral ties with Mercosur.

Energy integration is high on the government's agenda. Fidel Castro was invited to the inauguration of an electricity link between southern Venezuela and northern Brazil in 2001, and there are plans to build a natural gas connection with Colombia. On the oil side, Caracas recently revived strategic ties with fellow oil producers Colombia and Mexico, in the so-called "Group of the Three".

Venezuela is also actively promoting oil cooperation with Central America and the Caribbean. In 1980, Venezuela and Mexico established the San José Accords, with the objective of preventing disruption to the trade balance of the poorer countries in the region. Under this agreement, the two countries jointly supply 11 Central America and Caribbean countries with 160,000 barrels per day (bpd) of crude oil on preferential credit terms. In October 2000, the Venezuelan government set up, independently of Mexico, another oil co-operation agreement designed to supplement the San José Accords and relieve countries from high oil prices. Under this so-called Caracas Energy Accord, ten Caribbean and Central American countries receive 80,000 bpd at favourable credit terms when the oil price is more than US$15 per barrel. A separate energy agreement was established with Cuba.

## OVERVIEW OF THE VENEZUELAN ENERGY SECTOR

Venezuela is the world's sixth-largest producer and the fourth-largest exporter of crude oil. In 2001, it produced 3.1 million barrels per day (mb/d) of crude oil and natural gas liquids (NGL). Oil and gas dominate Venezuela's primary energy mix, accounting for 44% and 46% of total primary energy requirements respectively. Venezuela also has large hydropower resources, covering about 75% of the country's electricity needs, and the second-largest coal reserves in South America, though virtually all coal (5.8 Mtoe in 2000) is exported.

Venezuela's total primary energy use (TPES) was 59.3 Mtoe in 2000, up 5.2% from 1999. From 1971 to 2000, TPES grew at an average annual rate of 3.8%, although growth rates fluctuated widely from year to year, reflecting macroeconomic performance and international oil prices. This is a lower growth rate than the average of 5.5% for South America.

Venezuela's per capita energy use is the highest in South America (excluding Trinidad & Tobago), standing at 2.4 toe per person in 2000. However, this is just half of the average OECD level (see comparative tables in *Annex 3*). Energy intensity, expressed as primary energy use per unit of GDP adjusted for purchasing power parity (PPP), is also the highest in South America after Trinidad & Tobago, and about twice the average for OECD countries. Energy intensity has been flat over the last ten years, after rising steadily through the 1970s and early 1980s. Due to the large share of hydrocarbons in its energy mix, carbon intensity, measured as tons of $CO_2$ emitted per unit of GDP adjusted for PPP, is the second-highest in South America, after Trinidad & Tobago.

**Figure 8.1** Primary energy supply in Venezuela by fuel, 1975-2000

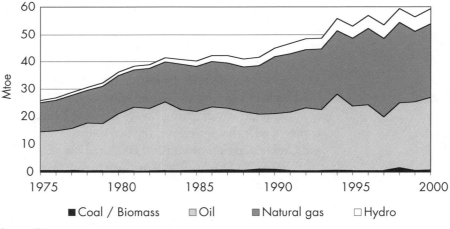

*Source: IEA.*

**Energy-sector policy and institutions**

In a dramatic turnaround from the previous government's *Apertura* policy, which had promoted foreign investment to help the country ramp up its oil production rapidly, the Chávez administration has reaffirmed strict state control over oil production and has made natural gas the cornerstone of its efforts to attract private investment to its energy sector.

Under the previous government, Venezuela's oil and gas sector was transformed, with reforms and decentralisation of all subsectors of the industry, from production to

**Figure 8.2** Electricity generation in Venezuela, 1975-2000

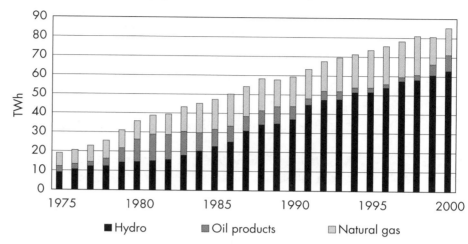

*Source: IEA.*

transportation, refining and marketing. From 1993 to 1998, state-owned company PDVSA established strategic alliances, operational agreements and profit-sharing joint ventures with several international oil companies.

Since President Chávez came to power in December 1998, energy policy has changed radically. He contends that the *Apertura* has benefited foreign companies at the expense of Venezuela's people and businesses. While respecting all contracts signed by the previous government, Chávez has made his energy-policy priorities clear from the start:

■ adhering strictly to OPEC production quotas and broader relations with OPEC countries in terms of technology exchange, commercial transactions, joint ventures, etc.;

■ retaining full state control over oil production, while promoting private-sector investment in natural gas and petrochemicals, but also in bitumen (Orimulsion) and coal;

■ increasing exports of refined products to 60% of total oil exports (crude and products) from the current 40%;

■ increasing the oil and gas reserve base, with a special focus on light and medium crude oil and non-associated gas.

The stated objective of the new administration's energy policy is not only to attract foreign capital to develop the country's natural gas industry, but also and most importantly, to encourage strong involvement of the domestic private sector. The aim is to make natural gas the foundation of a new wave of economic development, generating new non oil-related revenues and creating much-needed jobs.

The radical tones of Chávez's electoral campaign and initial declarations had created concern amongst foreign oil and gas investors. But since he took office, Chávez has taken a more pragmatic position. The new 1999 Constitution is viewed by some experts as more favourable to private oil and gas investors than the constitution of 1961, although some issues are subject to contradictory legal interpretation and will need to be clarified by the new organic hydrocarbon law.

The Ministry of Energy and Mines (MEM) is the government body responsible for energy policy and planning and for controlling all activities linked to hydrocarbons. With respect to natural gas, the MEM is responsible for:

- granting licences for exploration and development of non-associated gas reserves;

- controlling the fulfilment of all commitments established in the licences and sanctioning non-fulfilment;

- regulating the prices of gas, as well as the transmission and distribution tariffs;

- settling conflicts relating to the open-access regime.

The 1999 Gas Law created an autonomous regulatory agency, the *Ente Nacional del Gas* (Enagas), responsible for the regulation and promotion of competition in transportation and distribution.

*Petroleos de Venezuela S.A.* (PDVSA) is the state-owned oil and gas company. PDVSA's predecessor, Corporación Venezolana del Petróleo, was nationalised in 1975. The 1975 Nationalisation Law reserved all hydrocarbon activities to PDVSA and its affiliates, and stated that these affiliates must be 100% state-owned. The 1999 Constitution (Article 303) reaffirmed that PDVSA's shares are to remain in public hands, but, interestingly, this does not apply to PDVSA's affiliates and mixed companies.

The 1999 Constitution (Article 302) confirms that all "petroleum activity" shall be "reserved" to the state, through "appropriate Organic Laws and for reason of national interest". However, the word "reserved" does not necessarily bar PDVSA from carrying out business through agreements with private parties.[2] The Organic Hydrocarbons Law, enacted in November 2001, brought much-needed clarification and coherence to the maze of laws, regulations and explicit or tacit amendments and exceptions that currently govern the sector.

In any event, the word "petroleum" in the new constitution refers to crude oil and condensates, but not to gaseous hydrocarbons. The 1999 Gas Law[3] confirmed this, allowing private companies to operate in all segments of the gas chain except exploration and production of gas associated with oil.

In 1997, following the restructuring of PDVSA, the fully-owned subsidiary PDVSA Gas S.A. was set up to manage PDVSA's natural gas business from the point of delivery

2. In the past, despite this "reservation", the 1973 Law on the Internal Market had already opened some downstream activities to private companies. Then, during the 1990s, PDVSA was allowed to enter into operational agreements and ventures with foreign oil companies to develop marginal oil fields and ultra-heavy oil resources.
3. Organic Law of Gaseous Hydrocarbons, Decree No.310 of 12 September 1999, with rank and force of law.

by the producer to the end-user. In mid-1999, PDVSA created a new Gas Division within its Oil & Gas unit, to which PDVSA Gas will report. The new division is responsible for developing and planning business opportunities along the supply chain. It will promote and eventually be involved in projects to explore and produce non-associated gas, as well as other gas-related activities, such as the gathering, processing, transportation, industrialisation, distribution and domestic and international marketing of natural gas and its by-products. Over the course of time, PDVSA Gas has shifted emphasis from operations to marketing.

Since Chávez took office, changes in board members and high-level managers at PDVSA, as well as the restructuring of the company's activities, have given the government a much tighter grip over PDVSA than in the past.

## NATURAL GAS RESERVES AND PRODUCTION

Venezuela has the eighth-largest proven gas reserves in the world, standing at 4,163 billion cubic metres (bcm) as of 1 January 2002. Probable gas reserves in other frontier areas, such as the Orinoco Delta and off Venezuela's northern coast, are estimated at over 1,200 bcm. The 2000 US Geological Survey (USGS) evaluated undiscovered natural gas resources in Venezuela at over 2,800 bcm.

Although Venezuela has 58% of South America's total gas reserves, it accounts for only 37% of the continent's gross gas production, and for an even smaller proportion (27%) of the region's marketed gas production. The United States, which has a similar level of reserves (4,845 bcm), produces about ten times as much as Venezuela in terms of gross production, and has 20 times as much in terms of marketed production. This is partly due to the fact that much of the gas produced in Venezuela is reinjected into oil fields to maintain reservoir pressure and boost declining oil production.

In addition, 91% of Venezuela's gas reserves are associated with oil, and at present all gas production comes from associated oil and gas wells. Thus, gas output is highly dependent on oil output and hence on OPEC production quotas and trends in international oil prices. As shown in *Figure 8.3*, gross natural gas production and the share of reinjected gas are tightly correlated with the level of oil production.

In 2001, total gross gas production amounted to 62.4 bcm, of which 34% was reinjected. Some gas (5% in 2001) is also flared or vented, for lack of infrastructure to bring it to market. Volume shrinkage accounted for another 10% of gross production in 2001. Volume shrinkage is essentially the result of gas purification and/or extraction of natural gas's liquid fractions (ethane, LPG, natural gasoline and condensates). As a result, marketed production in 2001 was just half the level of gross production, at 29.3 bcm. At current levels of production, proven reserves are expected to last for 101 years.

Venezuela's sedimentary basins have historically been explored for oil. Thus there has been relatively little exploration in areas thought to be rich in non-associated gas. Gas

**Figure 8.3** Natural gas gross and marketed production, 1975-2001

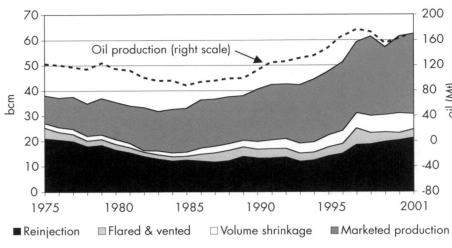

Sources: Cedigaz, IEA.

reserves are found in six main basins (see *Map 12* at the end of the book). The Oriental, Barinas-Guarico and Paria basins are the richest. Less abundant reserves are found in the Maracaibo, Barrancas and Yucal-Placer basins. About 85% of proven reserves are located in 18 giant fields, most of which were discovered before 1960. Gas production comes largely from oil fields in the Oriental, Maracaibo and Barinas-Guarico basins. As of yet, there is no production from the gas-rich Paria Basin. The Barinas-Guarino Basin is still underexplored in comparison to the Oriental and Maracaibo basins, though it is thought to contain most of Venezuela's non-associated gas.

PDVSA is currently the only gas producer in Venezuela. Although private companies are now allowed to explore and produce non-associated gas, PDVSA is likely to remain the largest gas producer for the foreseeable future.

To mitigate dependence on oil production and to increase and stabilise gas availability for end-users and exports, the government has launched an aggressive non-associated gas exploration and production programme, inviting the participation of private capital. The first-ever non-associated gas E&P licences were awarded in June 2001 to consortia including four foreign and two local companies (see later on in the chapter).

Fuller exploitation of associated gas reserves also features high on PDVSA's 2001-2006 Business Plan. PDVSA will not only increase gas output from its own fields, but will enter into agreements with private oil operators to encourage them to produce more gas from their marginal oil fields. Three such deals were signed in the first half of 2001 with three foreign investors who operate marginal oil fields: Argentina's Tecpetrol (Colon area), Japan's Teikoku Oil (East Guarico) and Spain's Repsol-YPF (Quiriquine). Together, these fields could supply 8.5 million cubic metres per day (mcm/d) as early as the end of 2003. Within three years, PDVSA expects other marginal oil fields in eastern Venezuela to add another 7 mcm/d.

Further ahead, an oil exploration contract held by Argentina's Perez Companc and another in La Vela, held by Phillips Petroleum, could be converted into gas licences to encourage gas production in the San Carlos field. However, the current terms of those contracts make gas production unprofitable because of the 67% income tax compared to the 34% offered to new gas licensees.

## The non-associated gas bidding round

After being postponed twice since it was first announced in early 2000, the first Venezuelan bidding round for non-associated gas areas finally took place at the end of June 2001, with only moderate success. Only six of the 11 areas on offer received bids, and only one received more than one bid. Of the companies or consortia that prequalified for the bidding (22 foreign and 11 domestic), only seven ultimately participated in the auction.

All of the 11 areas offered are located onshore in central and western Venezuela, covering an area of 13,500 square kilometres with estimated reserves of between 230-650 bcm. The cherries on the cake of the Venezuelan gas auction were the two Yucal Placer blocks. They have proven reserves of some 56 bcm (2 tcf) and lie just 100 kilometres southeast of Caracas, close to existing pipelines. Discovered in the 1960s, the Yucal field was exploited at a low rate from 1950 to 1989, but closed down when giant oil fields with associated gas were found in eastern Venezuela.

A consortium led by French TotalFinaElf (69.5%) and including Spanish Repsol-YPF (15%) as well as two local firms, Inepetrol (10.2%) and Otepi (5.3%), secured both the Yucal Placer blocks. The so-called Trio Consortium had to bid high on the Yucal

**Table 8.1** Blocks offered in Venezuela's 2001 gas auction

| Areas | No. Licences | Blocks | Type of Licence | Surface (km²) | Location | Estimated reserves* | Market |
|---|---|---|---|---|---|---|---|
| **Yucal Placer** | 2 | Yucal Placer Sur, Yucal Placer Norte | Proven Reserves | 1,811 | Guárico State, in center Venezuela | Proven 2 tcf Possible 3 tcf | Domestic market, Central/East Region |
| **Guarico-Cojedes** | 7 | Barbacoas, El Pao, El Totumo, La Galera, Memo, Tinaco, Tiznado | Exploration | 9,200 | North-Central Venezuela, across the states of Guárico (59%), Cojedes (27%), Aragua (14%) | Estimated 2-8 tcf | Domestic market, Central/East Region |
| **Barrancas** | 1 | Barrancas | Exploration | 1,970 | Western Venezuela across the states of Portuguesa (60%), Barinas (32%) and Trujillo (8%) | Estimated 2-6 tcf | Domestic market, Western Region |
| **Norte de Ambrosio** | 1 | Norte de Ambrosio | Exploration | 503 | Zulia State, in western Venezuela | Estimated 2-6 tcf | Domestic market, Western Region |
| **TOTAL** | 11 | | | 13,484 | | Est. 9-23 tcf | |

\* 1 tcf = 28.3 bcm.
Source: MEM.

Placer South to beat out Japan's Teikoku Oil. All other blocks received only a single bid each.

Repsol-YPF, bidding alone, also won the Barrancas block in the state of Barinas in south-west Venezuela. This is a block of 1,970 square kilometres. Substantial seismic analysis has already been carried out and three exploratory wells have been drilled. It is estimated to contain 56-170 bcm (2-6 tcf) of gas.

All the other blocks are in little-explored areas with no infrastructure in place. Seven blocks lie in the Guarico-Cojedes region, adjacent to the Yucal Placer area in south-central Venezuela. Argentina's Pluspetrol won the right to develop the little-explored Tinazdo and Barbacoas blocks, while fellow Argentine Perez Companc won the licence for the Tinaco block, located close to its San Carlos oil concession, where gas has also been found. The last area, Norte de Ambrosio, is in oil-rich Zulia in western Venezuela.

Award of the licences was based on the highest royalty offered over a minimum of 20%, instead of the usual system based on the highest signatory bonus. This system, known as *contraprestaciones,* was thought to be more attractive to potential bidders, because it does not involve an up-front lump-sum payment. The Venezuelan government initially required that at least one local company be part of each bidding consortium, but subsequently dropped that requirement for all blocks except Placer North and South. *Table 8.2* shows the royalty bonuses offered by the winning bidders.

**Table 8.2** Results of the 2001 Venezuelan gas auction

| Block | Winner | Royalty premium (over 20%) | Competitive bids |
| --- | --- | --- | --- |
| Yucal Placer Sur | Trio Consortium* | + 12.5% | Teikoku Oil |
| Yucal Placer Norte | Trio Consortium* | + 2.5% | None |
| Barrancas | Repsol YPF | + 2.59% | None |
| Barbacoas | Pluspetrol | + 1.5% | None |
| Tiznado | Pluspetrol | + 3.21% | None |
| Tinaco | Perez Companc | + 3.21% | None |
| 5 Other blocks | no bids | | |

* *TotalFinaElf (69.5%), Repsol YPF (15%), Inelectra (10.2%), Otepi (5.3%).*

Winning bidders have the right to explore, develop and produce gas for 35 years, with the possibility of a 30-year extension. They are required to develop at least 50% of the recoverable reserves of a given field. Producers will be allowed to sell their gas to any buyer in the market, with PDVSA Gas acting a "last-resort" buyer for the first seven years. However, the licences must indicate for what the gas produced will be used, and the Gas Law clearly stipulates that "all activities related to gaseous hydrocarbons shall be primarily related to national development...".

PDVSA expects that the first gas could flow from the Yucal Placer field very soon, followed by gas from others fields in 2006. The Yucal Placer consortium has announced that it plans to invest US$380 million over eight years, to produce 2 mcm/d by 2004 and 8.5 mcm/d by 2008. The other four licensed areas are expected to require another US$300 to 400 million of investments, most of them between 2006 and 2010.

TotalFinaElf, which claims to be the largest foreign investor in the country, was interested in securing "a strategic place at the forefront of the Venezuelan gas market as it opens to the private sector".

In February 2001, Repsol-YPF, which has the largest gas reserves in South America, signed an agreement with PDVSA to produce 30 bcm of associated gas from the mature Quiriquine oil block in Monagas. Similarly, the Argentine companies Pluspetrol and Perez Companc already have various interests in the Venezuelan oil sector. Japan's Teikoku Oil, too, has been awarded the right to produce natural gas from its East Guarico marginal field.

## NATURAL GAS DEMAND

Total primary supply of gas[4] was an estimated 29.3 bcm in 2001, not significantly higher from the 2000 level of 28.4 bcm and down from the 1998 level of 31.0 bcm. Primary gas supply grew at 3.8% annually from 1971 to 2000. During that period, the share of gas in total TPES remained fairly constant at 40-50%, with fluctuations reflecting trends in oil production.

**Figure 8.4** Natural gas demand in Venezuela, 1975-2000

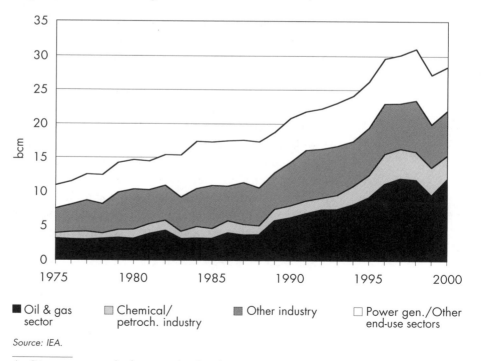

■ Oil & gas sector     ▨ Chemical/ petroch. industry     ▨ Other industry     □ Power gen./Other end-use sectors

*Source: IEA.*

---

4.  Primary energy supply of gas = marketed production + imports - exports.
    Marketed production = gross production - (reinjection + gas flared and vented + volume shrinkage).

The Venezuelan gas market is the world's 19th largest in terms of volume, but both production and consumption are highly concentrated in the hands of just a few players. PDVSA is the only producer and the largest consumer. Aside from the gas that PDVSA uses for reinjection and for conversion to liquids, which is not included in primary supply, it also uses natural gas to fuel its oil and gas extraction facilities and refineries. In addition, Pequiven, the PDVSA affiliate that dominates the petrochemical market, is the biggest industrial user of gas.

In 2000, 42% of gas supply was used by the petroleum industry, 17.5% in power generation, 35% by the industrial sector (mainly by the petrochemical and iron and steel industries) and only 4.8% by residential and commercial customers. Geographically, most of the demand is centred in and around the cities of Caracas and Maracaibo, since distribution networks are limited and these regions are close to the oil and gas fields and are the industrial heartland of the country.

The potential exists to increase gas consumption in all sectors. In the power-generation sector, Venezuela's highly-developed hydropower infrastructure had previously limited the scope for gas-fired generation, but this is now changing. The high share of hydropower in electricity generation renders the system vulnerable in case of droughts, and a larger share of thermal generation is seen as desirable to ensure security of supply. Moreover, hydropower capacity is concentrated in the southeast of the country, far from the main demand centres. The chronic lack of investment in maintenance and expansion of the electricity transmission system has produced severe bottlenecks, resulting in unreliable supply. This situation has created an incentive for industrial users situated close to oil and gas fields to turn to gas-fired auto-generation. In addition, several oil-fired plants are being converted to gas.

In the residential and commercial sectors, higher gas penetration will depend on the expansion of the distribution networks. Currently, only the cities of Caracas, Maracaibo, Puerto La Cruz, Barcelo and El Tigre have gas distribution networks that supply residential and commercial customers. Only 12% of Venezuelans use gas in their homes. As there is no need for space heating in Venezuela, residential and commercial demand will likely remain limited. At present, most households use bottled liquid petroleum gas (LPG) for cooking and electricity for water heating, even where a gas distribution network exists. The Ministry of Energy and Mines (MEM), which sets tariffs for all energy products, has set low tariffs for natural gas as an incentive for consumers to switch to natural gas. The government would like to promote new uses of gas, such as for air conditioning and refrigeration. It is also pushing forward an aggressive natural-gas vehicle (NGV) programme, aimed at substituting compressed natural gas for gasoline and diesel, though it is not clear how this programme will fit with the existing LPG-vehicle programme, launched in 1988.[5]

In the medium term, industry – in particular the petrochemical industry – is likely to remain the major gas user. Gas use in the petrochemical industry has grown 9% per year over the last decade. Pequiven, the PDVSA affiliate that dominates the petrochemical market, envisages trebling its gross sales in the next ten years. Other

---

5.  The country already has 18,000 vehicles running on LPG, out of a total of two million vehicles.

large industrial users of natural gas are the steel, aluminium and cement industries, but their gas demand has been flat over the past five years, owing to low industrial growth. Future gas demand in these sectors will depend crucially on the recovery of economic growth and investor confidence.

The MEM estimates that by 2008 natural gas could replace 7% of gasoline, 56% of diesel, 100% of kerosene and fuel oil and 64% of LPG, i.e. about 27% of total demand for oil products. Official projections see gas demand (excluding the petroleum-sector consumption) growing from the current 50 mcm/d to 120 mcm/d by 2008 (18 bcm/year to 44 bcm/year). Given current events, these projections seem a little too optimistic.

**Gas transmission and distribution**

Venezuela has about 5,000 kilometres of natural gas transmission and distribution pipelines, with a total capacity of 72 mcm/d. As shown on *Map 13* at the end of the book, there are two unconnected gas-transmission systems: the central-eastern system – which links Anaco with Puerto Ordaz, Puerto La Cruz, Caracas and Barquisimeto – and the western system around Lake Maracaibo.

The development of gas infrastructure in Venezuela has focused mainly on supplying the energy needs of the industrial sector. Most of the largest cities are connected to the main transmission lines, but only Caracas and adjacent areas, Maracaibo in the west, and Barcelona, Puerto La Cruz and El Tigre in the east, have distribution networks to supply residential and commercial consumers. As of the end of 1999, there were 1,500 industrial gas users and around half a million residential and commercial gas consumers.

The MEM estimates that over US$2 billion will be needed over the next decade to build 1,300 kilometres of new pipelines and expand the distribution network. Much of the new infrastructure is expected to be built with the participation of private investors. Projects currently envisaged include: a pipeline from Anaco in eastern Venezuela to Margarita Island; the doubling of the pipeline to Puerto Ordaz; and a new trunk line from San Felipe to Coro, connecting the eastern and western systems.

*PDVSA Gas* is PDVSA's division responsible for processing, transporting, distributing and marketing natural gas in the country. PDVSA Gas was created in 1997, regrouping the gas activities of the old PDVSA affiliates Corpoven, Maraven and Lagoven. PDVSA Gas operates the whole transportation network and some 1,700 kilometres of distribution pipelines in the Caracas and neighbouring areas. PDVSA gas accounts for nearly 100% of the industrial gas market and for 47% of the residential and commercial gas market.

There are three other distribution companies: *Venezolana Domestica de Gas* (DOMEGAS S.A.), *Venezolana Distribuidora de Gas Natural* (VDGAS S.A.) and *Fundación del Instituto Municipal de Energía* (FIME). FIME is the largest distributor, a non-profit entity that distributes and markets natural gas in Maracaibo and in the state of Zulia. It accounts for 43% of the country's residential and commercial gas market. The other two are private companies, which supply the Federal District and the eastern Anzoátegui State, and account for the remaining 10% share of the market.

# PROSPECTS FOR NATURAL GAS TRADE

**Piped gas trade**

Currently, Venezuela has no gas links with neighbouring countries. However, a project to build a cross-border pipeline between western Venezuela and Colombia is at an advanced stage of planning. The proposed pipeline would be approximately 200-km long and would connect the offshore fields and transport infrastructure of the Guajira region in northeast Colombia to markets in the Lake Maracaibo area in Venezuela. The pipeline is expected to start operating in 2005, supplying Colombian gas to Venezuela for a minimum of 7 years. The Lake Maracaibo area, where many of Venezuela's mature oil fields are concentrated, needs additional gas for reinjection, until Venezuelan non-associated gas production can be developed. In the longer term, the flow in the pipeline could be reversed to allow Venezuelan exports to Colombia, whose gas demand is expected to quickly outgrow domestic reserves.

Like other gas-rich countries in South America, Venezuela is also looking for opportunities to export some of its gas to Brazil's large and growing economy. However, the Brazilian states on Venezuela's southern border are among the poorest and least populated, and Brazil has discovered substantial gas reserves in the state of Amazonas, immediately south of Venezuela. The closest market for Venezuelan gas would be the densely populated and rapidly industrialising Northeast coast of Brazil, but over 3,500 kilometres of impenetrable forests and large rivers make exports of piped gas uneconomic. Although a pipeline link with Brazil is unlikely, LNG exports could be an option. There is a project to build an LNG regasification terminal on Brazil's Northeast coast. It is not clear if this project will go ahead, given recent reports of significant gas discoveries off the Northeast coast.

Venezuela also eyes the Central American market, which has no gas supply at the moment. However, Colombia is better placed, as it borders on Panama. Venezuelan gas would have to go through Colombia. In this respect, Venezuela's proposal for a cross-border pipeline to import gas from Colombia could in reality secure a route for future Venezuelan exports to Central America.

Venezuela is also exploring the possibility of exporting gas to the Caribbean islands. A project for a trans-Caribbean pipeline has long been on the drawing board, but with the costs of LNG transportation and regasification coming down, it now seems more likely that the Caribbean islands will import LNG rather than piped gas. In fact, several of them are already planning or constructing regasification facilities. In addition, Trinidad & Tobago, which has plans to increase significantly its LNG export capacity, is also looking at the Caribbean market and will likely get there well before Venezuela.

**LNG exports**

The main prospects for Venezuelan gas exports, whether to the Southern Cone or to North America and Europe, lie in LNG. However, if Venezuela wants to export LNG, it must first develop stable sources of non-associated gas, able to produce enough gas to feed at least two or three LNG trains (the cost of LNG from a single-train would probably be too high to compete with LNG from Trinidad & Tobago). After the weak response to the gas exploration licensing round in 2001, the government is now pushing ahead the development of two offshore areas in Eastern Venezuela thought to be rich in non-associated gas – the Deltana Platform and North Paria – with a view to export most of the gas that would be produced there as LNG.

In June 2002, Venezuela signed an agreement with Royal Dutch/Shell and Mitsubishi Corp. to develop natural gas production from an area north of the Paria peninsula, situated on Venezuela's northeastern coast in the Caribbean Sea. This project, which includes the construction of liquefied natural gas plants and terminal, is a downsized version of the old US$5-billion Cristobal Colón project, first conceived in 1989 and finally shelved in 2000 because of poor economics. The new scheme, known as Mariscal Sucre, is estimated to require some US$4.7 billion in investments. The project's partners are PDVSA (with a 60% stake), Royal Dutch/Shell (30%), Mitsubishi Corp. (8%) and Venezuelan investors (2%). PDVSA later signed a separate contract with Qatar Petroleum, which is expected to buy a 9%-10% stake from PDVSA. The project aims at producing some 30 mcm/d (1 bcf/d), beginning in 2007, of which 60%-70% would feed an initial one-train liquefaction plant. A second LNG train is also on the drawing board, which would take gas from both the North Paria and Deltana areas. Output from the first train is expected to be targeted exclusively at the US market, while LNG from the second train would be aimed at Europe and Brazil as well.

In 2002, Venezuela also started negotiations with companies interested in developing gas production in Plataforma Deltana, an area offshore the Orinoco delta, immediately south of Trinidad & Tobago's prolific gas fields. Five foreign companies were preselected in August 2002 to bid on four blocks: British BG and BP, US-based ChevronTexaco and ExxonMobil, French TotalFinaElf and Norwegian Statoil. ExxonMobil withdrew from the process at the last minute, but is said to be still negotiating for a fifth block. Overall, the development of the Deltana blocks are estimated to require about US$4 billion investment during the first six to eight years, and could produce another 30 mcm/d (1 bcf/d).

Given its geographical location, Venezuela has the potential to become a major LNG exporter to the Atlantic market, and in particular to the United States. But timing is crucial, as Venezuela will face stiff competition not only from neighbouring Trinidad & Tobago, whose LNG infrastructure is already in place and expanding rapidly, but also from other existing and projected LNG sources, all aiming at supplying the US East Coast (e.g. Nigeria).

In addition to LNG, PDVSA is also evaluating the technical and economic feasibility of a gas-to-liquids (GTL) project. The plant, to be located on the eastern Venezuelan coast, would produce a large share of high-quality diesel (high cetane, with no sulphur or aromatics) that could be sold as a final product or blended with low-cetane diesel, thereby enhancing the quality and quantity of diesel exports. The plant is expected to begin operation in 2007 and would produce around 10,000 bpd of products.

## GAS-SECTOR OPENING

**The 1999 Gas Law**

With the objective of boosting the development of Venezuela's large natural gas resources, an Organic Law on Gaseous Hydrocarbons (Gas Law) was passed in September 1999, replacing the 1971 Gas Law and opening up the sector to private investment. The government had initially envisaged that legislation on gaseous hydrocarbons would be included in a comprehensive Organic Hydrocarbons Law. But the Enabling Law of April 1999, which gave the president six months to pass laws on a number of issues without the approval of Congress, included gas but not oil.

The new law allows participation of private investors, national and foreign, in conjunction with state-owned entities or on their own, in the exploration and production of non-associated natural gas. It also allows private investment in the processing, storage, transportation, distribution and commercialisation of all gaseous hydrocarbons, including associated natural gas, natural gas liquids, liquid petroleum gas and refinery gas. The new law effectively ends PDVSA's monopoly in all areas of the gas chain, except in the exploration and production of associated gas, because of its links with oil production.

Natural gas resources remain the property of the state, and private companies operate under licences or permits granted by the MEM. Royalty is established at 20%. Under the new law, private companies are prohibited from participating in more than one part of the gas chain in each region. An exception to this rule may be authorised by the MEM, when vertical integration is the only viable form of a project. In this case, separate accounts have to be established for each activity.

Under the new law, licences and permits will be granted for specific projects approved by the MEM. The law states that projects aimed at expanding the use of gas as fuel for the domestic, residential, commercial and industrial sectors, or as raw material for high-value-added industrial production, shall take priority over export projects.

The law establishes open access to transmission, distribution and storage infrastructure, when there is available capacity. Conditions are to be negotiated by the parties. When they cannot agree, the MEM will intervene. The MEM also sets limits to the tariffs that will be paid by the final users.

Finally, the new law creates an autonomous regulatory agency, the *Ente Nacional del Gas* (Enagas), responsible for the regulation and promotion of competition in transportation and distribution. Its main tasks are:

■ to promote and supervise transmission, distribution, storage and supply activities;

■ to develop competition and ensure efficient use of the systems by, among other things, promoting the development of a secondary market for pipeline capacity;

■ to ensure open access to the systems and monitor anti-competitive and discriminatory behaviour;

■ to advise the MEM and the Ministry of Production and Commerce on the setting of distribution and transmission tariffs "while no true competition exists in the sector".

Enagas is governed by a board of five members: a president, a vice-president and three directors, all appointed by the MEM after consultation with the president of Venezuela. Board members are named for a three-year period, which can be renewed for successive periods. Although the new law aims at establishing an autonomous regulatory agency, the appointment of board members by the government, which presumably can remove them as well, may significantly limit the autonomy of Enagas.

**Natural-gas pricing policy**

The final piece of regulation that was needed before the first gas licensing round could go ahead was a ministerial resolution setting natural gas prices and tariffs. Two resolutions were issued in March 2001. One sets the maximum prices at which gas can be sold to different categories of customers. The other establishes maximum tariffs for the transportation and distribution of gas for the period 2002-2007.

Resolution No. 33 of 12 March 2001 establishes a scale of maximum gas prices for each type of consumer (residential and commercial customers, the petrochemical industry and other buyers) in two distribution areas: Anaco in the east and Lago in the west. Household customers will pay the lowest prices, and those prices will remain constant for the whole period 2002-2007. Household customers are defined by a maximum level of demand that varies according to the city. Gas prices for the petrochemical sector and other industry, initially set at the same level in each area, are allowed to rise, though in different ways. While the price for other industry rises according to a set scale, the price rises for the petrochemical industry will be determined by a formula to be set later by Enagas, which will take account of the efficiency of the petrochemical sector. All prices are based on the parity of the Venezuelan bolivar against the US dollar on

**Table 8.3** Natural gas prices in Venezuela, 2002-2007

| Bolivars/cubic metres* | 2001 | 2002 | 2003 | 2004 | 2005 | 2006 | 2007 |
|---|---|---|---|---|---|---|---|
| **Anaco Centre** | | | | | | | |
| Residential/Commercial | 9.9 | 9.9 | 9.9 | 9.9 | 9.9 | 9.9 | 9.9 |
| Petrochemicals | 12.4 | 13.6 | 13.6 | 13.6 | 13.6 | 13.6 | 13.6 |
| Other Industry | 12.4 | 13.6 | 14.8 | 16.3 | 18.0 | 20.0 | 22.2 |
| **Lago Centre** | | | | | | | |
| Residential/Commercial | 19.8 | 19.8 | 19.8 | 19.8 | 19.8 | 19.8 | 19.8 |
| Petrochemicals | 24.7 | 24.7 | 24.7 | 24.7 | 24.7 | 24.7 | 24.7 |
| Other Industry | 24.7 | 24.7 | 26.0 | 27.2 | 28.4 | 29.7 | 30.9 |

| US cents/million Btu** | 2001 | 2002 | 2003 | 2004 | 2005 | 2006 | 2007 |
|---|---|---|---|---|---|---|---|
| **Anaco Centre** | | | | | | | |
| Residential/Commercial | 40 | 40 | 40 | 40 | 40 | 40 | 40 |
| Petrochemicals | 50 | 55 | 55 | 55 | 55 | 55 | 55 |
| Other Industry | 50 | 55 | 60 | 66 | 74 | 82 | 91 |
| **Lago Centre** | | | | | | | |
| Residential/Commercial | 81 | 81 | 81 | 81 | 81 | 81 | 81 |
| Petrochemicals | 101 | 101 | 101 | 101 | 101 | 101 | 101 |
| Other Industry | 101 | 101 | 106 | 111 | 116 | 121 | 126 |

\*   As published in Resolution No. 33, 12 March 2001 (Source: MEM)
\*\* Calculated assuming 1 million Btu = 28.32 cubic metres and exchange rate: 695 Bolivars/US$ on 1 January 2001.

1 January 2001, and will be adjusted at the beginning of each year in line with the current exchange rate.

A transparent pricing system that offers a fair rate of return is critical if Venezuela is to attract large-scale foreign investment in the gas sector. The government claims that the maximum prices for sales on the local market and the royalty/taxation regime applicable to non-associated gas reserves guarantee investors an internal rate of return of around 15%. At the same time, by keeping end-user prices relatively low, the government aims to encourage a surge in gas demand. In particular, maximum prices for households are set at a very low level, substantially lower than previously. This will make gas competitive with other (subsidised) oil products used by households, but it is doubtful whether it will generate the investment needed to extend the distribution network. Another problem is that, since gas prices are all inclusive, they create considerable distortion, as distribution (and metering) for residential and commercials is specifically much more expensive than for large customers.

Critics point out that the pricing policy takes no account of inflation in the United States and other gas-consuming countries, nor of changes in exchange rate during the year. In addition, the pricing policy could result in PDVSA's buying natural gas at the wellhead at a price higher than the price it is allowed to charge to domestic consumers.

## CHALLENGES AND UNCERTAINTIES

The non-associated gas auction was the first opportunity to test investors' interest in Venezuela's oil and gas sector since Chávez came to power. Despite the government's optimistic declarations on the results of the licensing round, investors' response was actually rather subdued. Venezuela may well hold the continent's largest gas reserves and be ideally positioned to supply the Atlantic LNG market, but the conditions and safeguards offered to private investors and the marketing perspectives are probably not attractive enough to compete with the opportunities offered by other gas-rich countries.

Many factors explain foreign investor's scepticism and cloud Venezuela's gas prospects. One has to do with the characteristics of the acreage that was on offer in that round. The Venezuelan blocks paled in comparison with those offered, for example, in Saudi Arabia, which attracted strong interest among international companies. The Venezuelan blocks were small and relatively unexplored. More desirable areas farther to the east offered in 2002 attracted more interest. Enthusiasm may also have been tempered by the disappointing exploration results experienced by winners in Venezuela's previous oil-bidding rounds.

Even so, the main issue affecting investors' decisions to participate in the Venezuela licensing round was probably not the upstream risk, but rather the uncertainties of the gas marketing. The issue is less one of whether gas exists, but whether the gas can be sold and at what price. The government intends to give priority to the use of gas for oil recovery and domestic consumption over export projects. This is clearly demonstrated by the government's continued delay in giving the green light to the proposed LNG projects. Few investors will be interested in Venezuela's gas sector if they cannot sell gas on the export market. The ability to exploit export opportunities would not only reduce risk by providing market diversity, but would also protect investors against currency risks.

Timing is also a critical issue. LNG export capacity in neighbouring Trinidad & Tobago is increasing rapidly and competition to supply the gas-hungry US market is set to intensify with the entry of new players, such as Bolivia. If Venezuela does not refocus its priorities, interest by foreign investors in Venezuelan LNG facilities may wane in the coming years.

While the government emphasizes the opportunities that exist to increase supply to the domestic market, a sober analysis gives a less-than-bright outlook. Most additional gas demand is likely to come from the petroleum sector, where the need for reinjection is expected to grow substantially in the short-to-medium term. However, this need is likely to be supplied largely by PDVSA with associated gas. Non-petroleum gas demand has remained roughly flat since 1994. Additional gas demand from the industrial sector will depend on renewed economic growth and on the government's capacity to regain investor confidence through sound economic policies. Several large petrochemical projects might stimulate gas demand. But will producers be willing to sign long-term supply contracts at the low prices fixed for the sector? Finally, high reserve margins in the hydro-dominated power sector and the prospects for low gas use during years of ample hydro reserves will also slow the pace of gas-fired power development in Venezuela.

The key issue, and major source of concern, is the government's pricing policy. Venezuela has a long tradition of energy subsidies. The government's stated objective is to maintain end-user prices low enough to stimulate domestic demand. The government claims that the established price caps, combined with the royalty/taxation regime, will guarantee investors an internal rate of return of around 15%. While this may be true, investors may not find this internal rate of return adequate, given the uncertainties surrounding gas developments and business in general in Venezuela.

An additional risk concerns competition from associated gas. PDVSA's own exploration and production programme is likely to increase gas supply. Also, PDVSA is entering into agreements with private oil operators, under which they will produce more gas from their marginal oil fields or reopen gas-rich marginal fields, which were previously closed for lack of gas demand or infrastructure. Gas from these fields, which have already been fully explored, will be produced at a very low marginal cost and can easily displace the new, more expensive non-associated gas. Even if gas-to-gas competition can be ensured by 2007, which is by no means certain, gas will still have to compete with other fuels, whose supply and price are controlled by PDVSA.

Finally, investors remain wary of the country's economic and political uncertainties. Opposition to President Chávez is growing. Increasing government intervention in the energy sector has generated discontent in the business community and has led to an escalating dispute between PDVSA's managers and staff and the President, culminating in the ongoing general strike.

# ANNEX 1
# ENERGY BALANCES

# SOUTH AMERICA

## Energy Balance, 2000

| SUPPLY AND CONSUMPTION | Coal | Crude Oil | Petroleum Products | Gas | Nuclear | Hydro | Geotherm. Solar etc. | Combust. Renew. & Waste | Electricity | Heat | Total |
|---|---|---|---|---|---|---|---|---|---|---|---|
| | | | Thousand tonnes of oil equivalent (ktoe) | | | | | | | | |
| Production | 33648 | 362732 | – | 88814 | 3186 | 45904 | 59 | 61003 | – | – | 595348 |
| Imports | 15284 | 42749 | 21907 | 6008 | – | – | – | 5 | 4546 | – | 90498 |
| Exports | –29302 | –174931 | –65581 | –8440 | – | – | – | –120 | –4678 | – | –283050 |
| Intl. Marine Bunkers | – | – | –5559 | – | – | – | – | – | – | – | –5559 |
| Stock Changes | 1397 | –835 | –502 | – | – | – | – | 1265 | – | – | 1324 |
| **Total Primary Supply** | **21027** | **229717** | **–49734** | **86384** | **3186** | **45904** | **59** | **62154** | **–131** | **–** | **398564** |
| Transfers | – | –8156 | 8218 | – | – | – | – | – | – | – | 61 |
| Statistical Differences | 121 | –751 | –747 | –13 | – | – | – | 5 | 17 | – | –1371 |
| Electricity Plants | –6387 | –386 | –8884 | –20173 | –3186 | –45904 | –6 | –1398 | 60461 | – | –25865 |
| CHP Plants | – | – | – | – | – | – | – | –1309 | 573 | – | –737 |
| Heat Plants | – | – | – | – | – | – | – | – | – | – | – |
| Gas Works | – | – | –62 | –126 | – | – | – | – | – | – | –189 |
| Petroleum Refineries | – | –223012 | 217985 | – | – | – | – | – | – | – | –5027 |
| Coal Transformation | –4010 | – | –681 | – | – | – | – | – | – | – | –4691 |
| Liquefaction Plants | – | 3131 | 1892 | –6476 | – | – | – | – | – | – | –1454 |
| Other Transformation | – | – | – | – | – | – | – | –4026 | – | – | –4026 |
| Own Use | –623 | –145 | –10597 | –20074 | – | – | – | –5 | –1784 | – | –33228 |
| Distribution Losses | –657 | –71 | –353 | –748 | – | – | – | – | –10090 | – | –11921 |
| **Total Final Consumption** | **9468** | **328** | **157034** | **38769** | **–** | **–** | **53** | **55419** | **49040** | **–** | **310115** |
| **INDUSTRY SECTOR** | **9277** | **319** | **33977** | **26304** | **–** | **–** | **–** | **29666** | **22858** | **–** | **122401** |
| Iron and Steel | 5605 | 16 | 1239 | 5699 | – | – | – | 4718 | 2975 | – | 20253 |
| Chemical and Petrochemical | 174 | 72 | 15363 | 9905 | – | – | – | 1037 | 1733 | – | 28286 |
| of which: Feedstocks | – | 11 | 10469 | 4990 | – | – | – | 768 | – | – | 16237 |
| Non–Ferrous Metals | 244 | – | 1935 | 1267 | – | – | – | 8 | 4426 | – | 7881 |
| Non–Metallic Minerals | 1209 | 51 | 3512 | 2512 | – | – | – | 2093 | 784 | – | 10161 |
| Transport Equipment | – | – | – | 63 | – | – | – | – | – | – | 63 |
| Machinery | – | 23 | 233 | 4 | – | – | – | – | 44 | – | 304 |
| Mining and Quarrying | 406 | – | 1422 | 134 | – | – | – | – | 651 | – | 2613 |
| Food and Tobacco | 286 | 68 | 1319 | 1090 | – | – | – | 11811 | 1582 | – | 16154 |
| Paper Pulp and Printing | 397 | – | 1253 | 633 | – | – | – | 4853 | 1400 | – | 8535 |
| Wood and Wood Products | 1 | 33 | 2 | 53 | – | – | – | – | 14 | – | 104 |
| Construction | – | 10 | 49 | – | – | – | – | – | 5 | – | 64 |
| Textile and Leather | 216 | 29 | 323 | 272 | – | – | – | 82 | 749 | – | 1673 |
| Non–specified | 741 | 16 | 7328 | 4670 | – | – | – | 5064 | 8495 | – | 26314 |
| **TRANSPORT SECTOR** | **1** | **–** | **89110** | **2734** | **–** | **–** | **–** | **6145** | **198** | **–** | **98185** |
| Air | – | – | 7400 | – | – | – | – | – | – | – | 7400 |
| Road | – | – | 79426 | 1724 | – | – | – | 6138 | 11 | – | 87298 |
| Rail | 1 | – | 462 | – | – | – | – | – | 187 | – | 648 |
| Pipeline Transport | – | – | – | 890 | – | – | – | – | – | – | 890 |
| Internal Navigation | – | – | 1811 | – | – | – | – | – | – | – | 1811 |
| Non–specified | – | – | 12 | 120 | – | – | – | 7 | – | – | 139 |
| **OTHER SECTORS** | **80** | **8** | **24027** | **9733** | **–** | **–** | **53** | **19610** | **25988** | **–** | **79497** |
| Agriculture | 18 | 3 | 8377 | – | – | – | – | 2995 | 1337 | – | 12728 |
| Comm. and Publ. Services | – | 5 | 2425 | 2060 | – | – | – | 154 | 11247 | – | 15891 |
| Residential | 62 | – | 12919 | 7674 | – | – | 53 | 16443 | 13287 | – | 50436 |
| Non–specified | – | – | 305 | – | – | – | – | 18 | 115 | – | 439 |
| **NON–ENERGY USE** | **110** | **–** | **9924** | **–** | **–** | **–** | **–** | **–** | **–** | **–** | **10034** |
| in Industry/Transf./Energy | 110 | – | 9924 | – | – | – | – | – | – | – | 10034 |
| in Transport | – | – | – | – | – | – | – | – | – | – | – |
| in Other Sectors | – | – | – | – | – | – | – | – | – | – | – |
| **Electr. Generated – GWh** | **25920** | **–** | **35934** | **90821** | **12223** | **533789** | **73** | **10933** | **–** | **–** | **709693** |
| Electricity Plants | 25920 | – | 35934 | 90812 | 12223 | 533789 | 73 | 4284 | – | – | 703035 |
| CHP Plants | – | – | – | 9 | – | – | – | 6649 | – | – | 6658 |

Note: South America is defined here as including Argentina, Bolivia, Brazil, Chile, Colombia, Ecuador, Paraguay, Peru, Trinidad & Tobago, Uruguay and Venezuela.

# ARGENTINA

## Energy Balance, 2000

| SUPPLY AND CONSUMPTION | Coal | Crude Oil | Petroleum Products | Gas | Nuclear | Hydro | Geotherm. Solar etc. | Combust. Renew. & Waste | Electricity | Heat | Total |
|---|---|---|---|---|---|---|---|---|---|---|---|
| Production | 153 | 40158 | – | 33866 | 1610 | 2480 | 6 | 2947 | – | – | 81221 |
| Imports | 578 | 1355 | 1204 | 500 | – | – | – | – | 623 | – | 4261 |
| Exports | –220 | –14249 | –5066 | –3885 | – | – | – | – | –518 | – | –23938 |
| Intl. Marine Bunkers | – | – | –505 | – | – | – | – | – | – | – | –505 |
| Stock Changes | 17 | 134 | 280 | – | – | – | – | – | – | – | 431 |
| **Total Primary Supply** | **528** | **27398** | **–4087** | **30482** | **1610** | **2480** | **6** | **2947** | **106** | **–** | **61469** |
| Transfers | – | – | – | – | – | – | – | – | – | – | – |
| Statistical Differences | 72 | –39 | –1243 | 86 | – | – | – | 8 | – | – | –1117 |
| Electricity Plants | –470 | – | –909 | –9079 | –1610 | –2480 | –6 | –206 | 7655 | – | –7105 |
| CHP Plants | – | – | – | – | – | – | – | – | – | – | – |
| Heat Plants | – | – | – | – | – | – | – | – | – | – | – |
| Gas Works | – | – | – | – | – | – | – | – | – | – | – |
| Petroleum Refineries | – | –27359 | 26687 | – | – | – | – | – | – | – | –672 |
| Coal Transformation | 283 | – | –681 | – | – | – | – | – | – | – | –398 |
| Liquefaction Plants | – | – | 1771 | –1818 | – | – | – | – | – | – | –48 |
| Other Transformation | – | – | – | – | – | – | – | –145 | – | – | –145 |
| Own Use | –7 | – | –1202 | –3314 | – | – | – | – | –290 | – | –4813 |
| Distribution Losses | –17 | – | – | –254 | – | – | – | – | –980 | – | –1252 |
| **Total Final Consumption** | **388** | **–** | **20335** | **16102** | **–** | **–** | **–** | **2605** | **6490** | **–** | **45920** |
| **INDUSTRY SECTOR** | **388** | **–** | **2703** | **6568** | **–** | **–** | **–** | **2143** | **2996** | **–** | **14797** |
| Iron and Steel | 168 | – | – | 1394 | – | – | – | – | – | – | 1562 |
| Chemical and Petrochemical | – | – | 2101 | 1175 | – | – | – | – | – | – | 3276 |
| of which: Feedstocks | – | – | 2101 | 1175 | – | – | – | – | – | – | 3276 |
| Non–Ferrous Metals | – | – | – | 567 | – | – | – | – | – | – | 567 |
| Non–Metallic Minerals | – | – | – | 864 | – | – | – | – | – | – | 864 |
| Transport Equipment | – | – | – | 63 | – | – | – | – | – | – | 63 |
| Machinery | – | – | – | – | – | – | – | – | – | – | – |
| Mining and Quarrying | – | – | – | – | – | – | – | – | – | – | – |
| Food and Tobacco | – | – | – | 814 | – | – | – | – | – | – | 814 |
| Paper Pulp and Printing | – | – | – | 299 | – | – | – | – | – | – | 299 |
| Wood and Wood Products | – | – | – | 37 | – | – | – | – | – | – | 37 |
| Construction | – | – | – | – | – | – | – | – | – | – | – |
| Textile and Leather | – | – | – | 100 | – | – | – | – | – | – | 100 |
| Non–specified | 220 | – | 602 | 1256 | – | – | – | 2143 | 2996 | – | 7217 |
| **TRANSPORT SECTOR** | **–** | **–** | **12758** | **2289** | **–** | **–** | **–** | **–** | **45** | **–** | **15092** |
| Air | – | – | 1843 | – | – | – | – | – | – | – | 1843 |
| Road | – | – | 10859 | 1399 | – | – | – | – | – | – | 12258 |
| Rail | – | – | – | – | – | – | – | – | 45 | – | 45 |
| Pipeline Transport | – | – | – | 890 | – | – | – | – | – | – | 890 |
| Internal Navigation | – | – | 56 | – | – | – | – | – | – | – | 56 |
| Non–specified | – | – | – | – | – | – | – | – | – | – | – |
| **OTHER SECTORS** | **–** | **–** | **3796** | **7245** | **–** | **–** | **–** | **462** | **3450** | **–** | **14953** |
| Agriculture | – | – | 2436 | – | – | – | – | 48 | 45 | – | 2529 |
| Comm. and Publ. Services | – | – | 140 | 1431 | – | – | – | – | 1566 | – | 3138 |
| Residential | – | – | 1220 | 5814 | – | – | – | 414 | 1838 | – | 9286 |
| Non–specified | – | – | – | – | – | – | – | – | – | – | – |
| **NON–ENERGY USE** | **–** | **–** | **1078** | **–** | **–** | **–** | **–** | **–** | **–** | **–** | **1078** |
| in Industry/Transf./Energy | – | – | 1078 | – | – | – | – | – | – | – | 1078 |
| in Transport | – | – | – | – | – | – | – | – | – | – | – |
| in Other Sectors | – | – | – | – | – | – | – | – | – | – | – |
| **Electr. Generated – GWh** | **1562** | **–** | **3096** | **49056** | **6177** | **28841** | **71** | **211** | **–** | **–** | **89014** |
| Electricity Plants | 1562 | – | 3096 | 49056 | 6177 | 28841 | 71 | 211 | – | – | 89014 |
| CHP Plants | – | – | – | – | – | – | – | – | – | – | – |

Thousand tonnes of oil equivalent (ktoe)

*Source: IEA (2002) Energy Balances of Non–OECD Countries 1999–2000.*

# BOLIVIA

## Energy Balance, 2000

| SUPPLY AND CONSUMPTION | Coal | Crude Oil | Petroleum Products | Gas | Nuclear | Hydro | Geotherm. Solar etc. | Combust. Renew. & Waste | Electricity | Heat | Total |
|---|---|---|---|---|---|---|---|---|---|---|---|
| | | | | Thousand tonnes of oil equivalent (ktoe) | | | | | | | |
| Production | – | 1902 | – | 3106 | – | 170 | – | 723 | – | – | 5901 |
| Imports | – | – | 290 | – | – | – | – | – | 1 | – | 291 |
| Exports | – | – | – | –1775 | – | – | – | – | – | – | –1775 |
| Intl. Marine Bunkers | – | – | – | – | – | – | – | – | – | – | – |
| Stock Changes | – | 512 | – | – | – | – | – | – | – | – | 512 |
| **Total Primary Supply** | **–** | **2415** | **290** | **1331** | **–** | **170** | **–** | **723** | **1** | **–** | **4929** |
| Transfers | – | –309 | 338 | – | – | – | – | – | – | – | 28 |
| Statistical Differences | – | –324 | –397 | –123 | – | – | – | – | – | – | –844 |
| Electricity Plants | – | – | –31 | –463 | – | –170 | – | –31 | 340 | – | –355 |
| CHP Plants | – | – | – | – | – | – | – | – | – | – | – |
| Heat Plants | – | – | – | – | – | – | – | – | – | – | – |
| Gas Works | – | – | – | – | – | – | – | – | – | – | – |
| Petroleum Refineries | – | –1781 | 1607 | – | – | – | – | – | – | – | –174 |
| Coal Transformation | – | – | – | – | – | – | – | – | – | – | – |
| Liquefaction Plants | – | – | – | – | – | – | – | – | – | – | – |
| Other Transformation | – | – | – | – | – | – | – | –23 | – | – | –23 |
| Own Use | – | – | –230 | –174 | – | – | – | – | –2 | – | –406 |
| Distribution Losses | – | – | – | –219 | – | – | – | – | –61 | – | –280 |
| **Total Final Consumption** | **–** | **–** | **1577** | **352** | **–** | **–** | **–** | **669** | **277** | **–** | **2876** |
| **INDUSTRY SECTOR** | **–** | **–** | **20** | **320** | **–** | **–** | **–** | **421** | **88** | **–** | **849** |
| Iron and Steel | – | – | – | – | – | – | – | – | – | – | – |
| Chemical and Petrochemical | – | – | – | – | – | – | – | – | – | – | – |
| of which: Feedstocks | – | – | – | – | – | – | – | – | – | – | – |
| Non–Ferrous Metals | – | – | – | – | – | – | – | – | – | – | – |
| Non–Metallic Minerals | – | – | – | – | – | – | – | – | – | – | – |
| Transport Equipment | – | – | – | – | – | – | – | – | – | – | – |
| Machinery | – | – | – | – | – | – | – | – | – | – | – |
| Mining and Quarrying | – | – | – | – | – | – | – | – | – | – | – |
| Food and Tobacco | – | – | – | – | – | – | – | – | – | – | – |
| Paper Pulp and Printing | – | – | – | – | – | – | – | – | – | – | – |
| Wood and Wood Products | – | – | – | – | – | – | – | – | – | – | – |
| Construction | – | – | 4 | – | – | – | – | – | – | – | 4 |
| Textile and Leather | – | – | – | – | – | – | – | – | – | – | – |
| Non–specified | – | – | 16 | 320 | – | – | – | 421 | 88 | – | 845 |
| **TRANSPORT SECTOR** | **–** | **–** | **931** | **22** | **–** | **–** | **–** | **–** | **–** | **–** | **953** |
| Air | – | – | 133 | – | – | – | – | – | – | – | 133 |
| Road | – | – | 778 | – | – | – | – | – | – | – | 778 |
| Rail | – | – | 6 | – | – | – | – | – | – | – | 6 |
| Pipeline Transport | – | – | – | – | – | – | – | – | – | – | – |
| Internal Navigation | – | – | 13 | – | – | – | – | – | – | – | 13 |
| Non–specified | – | – | – | 22 | – | – | – | – | – | – | 22 |
| **OTHER SECTORS** | **–** | **–** | **614** | **10** | **–** | **–** | **–** | **248** | **189** | **–** | **1061** |
| Agriculture | – | – | 2 | – | – | – | – | – | – | – | 2 |
| Comm. and Publ. Services | – | – | – | 6 | – | – | – | – | 66 | – | 73 |
| Residential | – | – | 310 | 4 | – | – | – | 248 | 116 | – | 678 |
| Non–specified | – | – | 301 | – | – | – | – | – | 6 | – | 308 |
| **NON–ENERGY USE** | **–** | **–** | **13** | **–** | **–** | **–** | **–** | **–** | **–** | **–** | **13** |
| in Industry/Transf./Energy | – | – | 13 | – | – | – | – | – | – | – | 13 |
| in Transport | – | – | – | – | – | – | – | – | – | – | – |
| in Other Sectors | – | – | – | – | – | – | – | – | – | – | – |
| **Electr. Generated – GWh** | **–** | **–** | **100** | **1816** | **–** | **1973** | **–** | **63** | **–** | **–** | **3952** |
| Electricity Plants | – | – | 100 | 1816 | – | 1973 | – | 63 | – | – | 3952 |
| CHP Plants | – | – | – | – | – | – | – | – | – | – | – |

Source: IEA (2002), Energy Balances of Non-OECD Countries 1999-2000.

## BRAZIL

## Energy Balance, 2000

| SUPPLY AND CONSUMPTION | Coal | Crude Oil | Petroleum Products | Gas | Nuclear | Hydro | Geotherm. Solar etc. | Combust. Renew. & Waste | Electricity | Heat | Total |
|---|---|---|---|---|---|---|---|---|---|---|---|
| Production | 2541 | 64477 | – | 5658 | 1576 | 26203 | – | 41623 | – | – | 142078 |
| Imports | 10994 | 20423 | 14098 | 1804 | – | – | – | 4 | 3801 | – | 51124 |
| Exports | – | –1663 | –5934 | – | – | – | – | –120 | –1 | – | –7716 |
| Intl. Marine Bunkers | – | – | –2909 | – | – | – | – | – | – | – | –2909 |
| Stock Changes | 103 | –264 | –517 | – | – | – | – | 1265 | – | – | 588 |
| **Total Primary Supply** | **13639** | **82973** | **4739** | **7462** | **1576** | **26203** | **–** | **42773** | **3801** | **–** | **183165** |
| Transfers | – | – | 863 | – | – | – | – | – | – | – | 863 |
| Statistical Differences | 112 | –200 | 655 | 10 | – | – | – | 2 | – | – | 579 |
| Electricity Plants | –2962 | –373 | –3903 | –532 | –1576 | –26203 | – | –583 | 29455 | – | –6676 |
| CHP Plants | – | – | – | – | – | – | – | –1309 | 572 | – | –738 |
| Heat Plants | – | – | – | – | – | – | – | – | – | – | – |
| Gas Works | – | – | – | –10 | – | – | – | – | – | – | –10 |
| Petroleum Refineries | – | –82401 | 82260 | – | – | – | – | – | – | – | –141 |
| Coal Transformation | –3748 | – | – | – | – | – | – | – | – | – | –3748 |
| Liquefaction Plants | – | – | – | – | – | – | – | – | – | – | – |
| Other Transformation | – | – | – | – | – | – | – | –3613 | – | – | –3613 |
| Own Use | –580 | – | –3938 | –1971 | – | – | – | – | –899 | – | –7388 |
| Distribution Losses | –531 | – | –346 | –164 | – | – | – | – | –5463 | – | –6504 |
| **Total Final Consumption** | **5930** | **–** | **80330** | **4794** | **–** | **–** | **–** | **37269** | **27465** | **–** | **155789** |
| **INDUSTRY SECTOR** | **5820** | **–** | **21865** | **4309** | **–** | **–** | **–** | **22310** | **12606** | **–** | **66910** |
| Iron and Steel | 4644 | – | 493 | 693 | – | – | – | 4718 | 1816 | – | 12365 |
| Chemical and Petrochemical | 74 | – | 11619 | 1877 | – | – | – | 1008 | 1479 | – | 16058 |
| of which: Feedstocks | – | – | 8368 | 692 | – | – | – | 768 | – | – | 9827 |
| Non–Ferrous Metals | 244 | – | 1399 | 140 | – | – | – | 8 | 2493 | – | 4284 |
| Non–Metallic Minerals | 219 | – | 3234 | 293 | – | – | – | 2078 | 617 | – | 6441 |
| Transport Equipment | – | – | – | – | – | – | – | – | – | – | – |
| Machinery | – | – | – | – | – | – | – | – | – | – | – |
| Mining and Quarrying | 406 | – | 1139 | 134 | – | – | – | – | 632 | – | 2311 |
| Food and Tobacco | 50 | – | 1143 | 214 | – | – | – | 9999 | 1393 | – | 12799 |
| Paper Pulp and Printing | 75 | – | 1053 | 258 | – | – | – | 3834 | 1044 | – | 6263 |
| Wood and Wood Products | – | – | – | – | – | – | – | – | – | – | – |
| Construction | – | – | – | – | – | – | – | – | – | – | – |
| Textile and Leather | – | – | 275 | 162 | – | – | – | 81 | 600 | – | 1119 |
| Non–specified | 108 | – | 1509 | 537 | – | – | – | 584 | 2532 | – | 5270 |
| **TRANSPORT SECTOR** | **–** | **–** | **41704** | **261** | **–** | **–** | **–** | **6131** | **108** | **–** | **48203** |
| Air | – | – | 3227 | – | – | – | – | – | – | – | 3227 |
| Road | – | – | 37135 | 261 | – | – | – | 6131 | – | – | 43526 |
| Rail | – | – | 407 | – | – | – | – | – | 108 | – | 514 |
| Pipeline Transport | – | – | – | – | – | – | – | – | – | – | – |
| Internal Navigation | – | – | 936 | – | – | – | – | – | – | – | 936 |
| Non–specified | – | – | – | – | – | – | – | – | – | – | – |
| **OTHER SECTORS** | **–** | **–** | **12394** | **225** | **–** | **–** | **–** | **8829** | **14752** | **–** | **36199** |
| Agriculture | – | – | 4608 | – | – | – | – | 1643 | 1069 | – | 7320 |
| Comm. and Publ. Services | – | – | 1374 | 80 | – | – | – | 147 | 6503 | – | 8104 |
| Residential | – | – | 6413 | 145 | – | – | – | 7038 | 7180 | – | 20775 |
| Non–specified | – | – | – | – | – | – | – | – | – | – | – |
| **NON–ENERGY USE** | **110** | **–** | **4367** | **–** | **–** | **–** | **–** | **–** | **–** | **–** | **4477** |
| in Industry/Transf./Energy | 110 | – | 4367 | – | – | – | – | – | – | – | 4477 |
| in Transport | – | – | – | – | – | – | – | – | – | – | – |
| in Other Sectors | – | – | – | – | – | – | – | – | – | – | – |
| **Electr. Generated – GWh** | **10072** | **–** | **16889** | **2389** | **6046** | **304691** | **1** | **9065** | **–** | **–** | **349153** |
| Electricity Plants | 10072 | – | 16889 | 2389 | 6046 | 304691 | 1 | 2416 | – | – | 342504 |
| CHP Plants | – | – | – | – | – | – | – | 6649 | – | – | 6649 |

Source: IEA (2002), Energy Balances of Non-OECD Countries 1999-2000.

# CHILE

## Energy Balance, 2000

| SUPPLY AND CONSUMPTION | Coal | Crude Oil | Petroleum Products | Gas | Nuclear | Hydro | Geotherm. Solar etc. | Combust. Renew. & Waste | Electricity | Heat | Total |
|---|---|---|---|---|---|---|---|---|---|---|---|
| | | | Thousand tonnes of oil equivalent (ktoe) | | | | | | | | |
| Production | 249 | 427 | – | 1752 | – | 1641 | – | 4231 | – | – | 8299 |
| Imports | 3123 | 10505 | 1941 | 3674 | – | – | – | – | – | – | 19243 |
| Exports | –29 | –1608 | –650 | – | – | – | – | – | – | – | –2287 |
| Intl. Marine Bunkers | – | – | –656 | – | – | – | – | – | – | – | –656 |
| Stock Changes | –88 | –263 | 155 | – | – | – | – | – | – | – | –196 |
| **Total Primary Supply** | **3255** | **9061** | **790** | **5426** | **-** | **1641** | **-** | **4231** | **-** | **-** | **24403** |
| Transfers | – | –102 | 113 | – | – | – | – | – | – | – | 11 |
| Statistical Differences | –3 | 7 | –159 | – | – | – | – | – | – | – | –155 |
| Electricity Plants | –2266 | – | –300 | –1652 | – | –1641 | – | –304 | 3549 | – | –2614 |
| CHP Plants | – | – | – | – | – | – | – | – | – | – | – |
| Heat Plants | – | – | – | – | – | – | – | – | – | – | – |
| Gas Works | – | – | –44 | 39 | – | – | – | – | – | – | –6 |
| Petroleum Refineries | – | –10574 | 9520 | – | – | – | – | – | – | – | –1054 |
| Coal Transformation | –217 | – | – | – | – | – | – | – | – | – | –217 |
| Liquefaction Plants | – | 1643 | – | –2315 | – | – | – | – | – | – | –672 |
| Other Transformation | – | – | – | – | – | – | – | – | – | – | – |
| Own Use | –2 | – | –509 | –371 | – | – | – | –5 | –150 | – | –1037 |
| Distribution Losses | –52 | – | – | –110 | – | – | – | – | –251 | – | –413 |
| **Total Final Consumption** | **713** | **35** | **9411** | **1017** | **-** | **-** | **-** | **3922** | **3147** | **-** | **18245** |
| **INDUSTRY SECTOR** | **703** | **35** | **2051** | **671** | **-** | **-** | **-** | **1129** | **2195** | **-** | **6785** |
| Iron and Steel | 254 | – | 62 | 7 | – | – | – | – | 30 | – | 353 |
| Chemical and Petrochemical | – | 35 | 2 | 11 | – | – | – | – | 47 | – | 95 |
| of which: Feedstocks | – | – | – | – | – | – | – | – | – | – | – |
| Non–Ferrous Metals | – | – | 440 | 48 | – | – | – | – | 1073 | – | 1562 |
| Non–Metallic Minerals | 169 | – | 73 | – | – | – | – | – | 34 | – | 276 |
| Transport Equipment | – | – | – | – | – | – | – | – | – | – | – |
| Machinery | – | – | – | – | – | – | – | – | – | – | – |
| Mining and Quarrying | – | – | 283 | – | – | – | – | – | 19 | – | 302 |
| Food and Tobacco | 105 | – | 10 | – | – | – | – | – | 7 | – | 121 |
| Paper Pulp and Printing | 26 | – | 153 | 29 | – | – | – | 792 | 274 | – | 1274 |
| Wood and Wood Products | – | – | – | – | – | – | – | – | – | – | – |
| Construction | – | – | – | – | – | – | – | – | – | – | – |
| Textile and Leather | – | – | – | – | – | – | – | – | – | – | – |
| Non–specified | 150 | – | 1029 | 575 | – | – | – | 338 | 711 | – | 2802 |
| **TRANSPORT SECTOR** | **-** | **-** | **5948** | **8** | **-** | **-** | **-** | **-** | **19** | **-** | **5974** |
| Air | – | – | 582 | – | – | – | – | – | – | – | 582 |
| Road | – | – | 5059 | 8 | – | – | – | – | 11 | – | 5078 |
| Rail | – | – | 19 | – | – | – | – | – | 8 | – | 26 |
| Pipeline Transport | – | – | – | – | – | – | – | – | – | – | – |
| Internal Navigation | – | – | 289 | – | – | – | – | – | – | – | 289 |
| Non–specified | – | – | – | – | – | – | – | – | – | – | – |
| **OTHER SECTORS** | **10** | **-** | **1413** | **338** | **-** | **-** | **-** | **2792** | **933** | **-** | **5486** |
| Agriculture | 10 | – | 168 | – | – | – | – | – | 14 | – | 191 |
| Comm. and Publ. Services | – | – | 157 | 71 | – | – | – | – | 388 | – | 616 |
| Residential | – | – | 1088 | 267 | – | – | – | 2792 | 532 | – | 4678 |
| Non–specified | – | – | – | – | – | – | – | – | – | – | – |
| **NON–ENERGY USE** | **-** | **-** | **-** | **-** | **-** | **-** | **-** | **-** | **-** | **-** | **-** |
| in Industry/Transf./Energy | – | – | – | – | – | – | – | – | – | – | – |
| in Transport | – | – | – | – | – | – | – | – | – | – | – |
| in Other Sectors | – | – | – | – | – | – | – | – | – | – | – |
| **Electr. Generated – GWh** | **11123** | **-** | **1204** | **9054** | **-** | **19081** | **-** | **806** | **-** | **-** | **41268** |
| Electricity Plants | 11123 | – | 1204 | 9054 | – | 19081 | – | 806 | – | – | 41268 |
| CHP Plants | – | – | – | – | – | – | – | – | – | – | – |

Source: IEA (2002), Energy Balances of Non–OECD Countries 1999–2000.

# COLOMBIA

## Energy Balance, 2000

| SUPPLY AND CONSUMPTION | Coal | Crude Oil | Petroleum Products | Gas | Nuclear | Hydro | Geotherm. Solar etc. | Combust. Renew. & Waste | Electricity | Heat | Total |
|---|---|---|---|---|---|---|---|---|---|---|---|
| | | | | Thousand tonnes of oil equivalent (ktoe) | | | | | | | |
| Production | 24937 | 36169 | – | 5456 | – | 2758 | – | 5264 | – | – | 74584 |
| Imports | – | 197 | 268 | – | – | – | – | – | 7 | – | 471 |
| Exports | –23264 | –20023 | –3809 | – | – | – | – | – | –3 | – | –47099 |
| Intl. Marine Bunkers | – | – | –233 | – | – | – | – | – | – | – | –233 |
| Stock Changes | 1170 | –148 | 41 | – | – | – | – | – | – | – | 1062 |
| **Total Primary Supply** | **2843** | **16194** | **–3733** | **5456** | **–** | **2758** | **–** | **5264** | **3** | **–** | **28786** |
| Transfers | – | –443 | 130 | – | – | – | – | – | – | – | –313 |
| Statistical Differences | –78 | –101 | 110 | 2 | – | – | – | –7 | 51 | – | –24 |
| Electricity Plants | –590 | –13 | –15 | –1887 | – | –2758 | – | –135 | 3779 | – | –1619 |
| CHP Plants | – | – | – | – | – | – | – | – | 1 | – | 1 |
| Heat Plants | – | – | – | – | – | – | – | – | – | – | – |
| Gas Works | – | – | – | – | – | – | – | – | – | – | – |
| Petroleum Refineries | – | –15286 | 15510 | – | – | – | – | – | – | – | 224 |
| Coal Transformation | –228 | – | – | – | – | – | – | – | – | – | –228 |
| Liquefaction Plants | – | – | – | – | – | – | – | – | – | – | – |
| Other Transformation | – | – | – | – | – | – | – | –98 | – | – | –98 |
| Own Use | –34 | –69 | –516 | –1946 | – | – | – | – | –65 | – | –2629 |
| Distribution Losses | –57 | –1 | – | – | – | – | – | – | –902 | – | –961 |
| **Total Final Consumption** | **1856** | **282** | **11487** | **1624** | **–** | **–** | **–** | **5024** | **2867** | **–** | **23140** |
| **INDUSTRY SECTOR** | **1793** | **273** | **1205** | **976** | **–** | **–** | **–** | **1754** | **982** | **–** | **6983** |
| Iron and Steel | 360 | 16 | 44 | 27 | – | – | – | – | 180 | – | 627 |
| Chemical and Petrochemical | 100 | 26 | 388 | 411 | – | – | – | 29 | 155 | – | 1110 |
| of which: Feedstocks | – | – | – | – | – | – | – | – | – | – | – |
| Non–Ferrous Metals | – | – | – | – | – | – | – | – | – | – | – |
| Non–Metallic Minerals | 689 | 51 | 119 | 374 | – | – | – | 15 | 133 | – | 1381 |
| Transport Equipment | – | – | – | – | – | – | – | – | – | – | – |
| Machinery | – | 23 | 233 | 4 | – | – | – | – | 44 | – | 304 |
| Mining and Quarrying | – | – | – | – | – | – | – | – | – | – | – |
| Food and Tobacco | 131 | 68 | 166 | 62 | – | – | – | 1481 | 182 | – | 2089 |
| Paper Pulp and Printing | 296 | – | 47 | 47 | – | – | – | 227 | 82 | – | 699 |
| Wood and Wood Products | 1 | 33 | 2 | 16 | – | – | – | – | 14 | – | 67 |
| Construction | – | 10 | 45 | – | – | – | – | – | 5 | – | 60 |
| Textile and Leather | 216 | 29 | 48 | 10 | – | – | – | 1 | 149 | – | 454 |
| Non–specified | – | 16 | 114 | 25 | – | – | – | – | 37 | – | 192 |
| **TRANSPORT SECTOR** | **1** | **–** | **6879** | **56** | **–** | **–** | **–** | **–** | **4** | **–** | **6940** |
| Air | – | – | 680 | – | – | – | – | – | – | – | 680 |
| Road | – | – | 6021 | 56 | – | – | – | – | – | – | 6077 |
| Rail | 1 | – | 19 | – | – | – | – | – | 4 | – | 24 |
| Pipeline Transport | – | – | – | – | – | – | – | – | – | – | – |
| Internal Navigation | – | – | 158 | – | – | – | – | – | – | – | 158 |
| Non–specified | – | – | – | – | – | – | – | – | – | – | – |
| **OTHER SECTORS** | **62** | **8** | **1596** | **593** | **–** | **–** | **–** | **3271** | **1881** | **–** | **7410** |
| Agriculture | – | 3 | 476 | – | – | – | – | 1157 | 111 | – | 1747 |
| Comm. and Publ. Services | – | 5 | 315 | 89 | – | – | – | – | 783 | – | 1192 |
| Residential | 62 | – | 805 | 504 | – | – | – | 2096 | 957 | – | 4424 |
| Non–specified | – | – | – | – | – | – | – | 18 | 30 | – | 48 |
| **NON–ENERGY USE** | **–** | **–** | **1808** | **–** | **–** | **–** | **–** | **–** | **–** | **–** | **1808** |
| in Industry/Transf./Energy | – | – | 1808 | – | – | – | – | – | – | – | 1808 |
| in Transport | – | – | – | – | – | – | – | – | – | – | – |
| in Other Sectors | – | – | – | – | – | – | – | – | – | – | – |
| **Electr. Generated – GWh** | **2955** | **–** | **101** | **8272** | **–** | **32074** | **–** | **550** | **–** | **–** | **43952** |
| *Electricity Plants* | *2955* | *–* | *101* | *8263* | *–* | *32074* | *–* | *550* | *–* | *–* | *43943* |
| *CHP Plants* | *–* | *–* | *–* | *9* | *–* | *–* | *–* | *–* | *–* | *–* | *9* |

*Source: IEA (2002), Energy Balances of Non–OECD Countries 1999–2000.*

# ECUADOR

## Energy Balance, 2000

| SUPPLY AND CONSUMPTION | Coal | Crude Oil | Petroleum Products | Gas | Nuclear | Hydro | Geotherm. Solar etc. | Combust. Renew. & Waste | Electricity | Heat | Total |
|---|---|---|---|---|---|---|---|---|---|---|---|
| *Thousand tonnes of oil equivalent (ktoe)* | | | | | | | | | | | |
| Production | – | 20887 | – | 281 | – | 654 | – | 697 | – | – | 22520 |
| Imports | – | – | 1039 | – | – | – | – | – | – | – | 1039 |
| Exports | – | –12366 | –2251 | – | – | – | – | – | – | – | –14617 |
| Intl. Marine Bunkers | – | – | –272 | – | – | – | – | – | – | – | –272 |
| Stock Changes | – | –76 | –408 | – | – | – | – | – | – | – | –484 |
| **Total Primary Supply** | **–** | **8445** | **–1891** | **281** | **–** | **654** | **–** | **697** | **–** | **–** | **8187** |
| Transfers | – | – | – | – | – | – | – | – | – | – | – |
| Statistical Differences | – | –431 | –63 | – | – | – | – | – | –26 | – | –521 |
| Electricity Plants | – | – | –713 | – | – | –654 | – | – | 913 | – | –454 |
| CHP Plants | – | – | – | – | – | – | – | – | – | – | – |
| Heat Plants | – | – | – | – | – | – | – | – | – | – | – |
| Gas Works | – | – | – | – | – | – | – | – | – | – | – |
| Petroleum Refineries | – | –7870 | 7696 | – | – | – | – | – | – | – | –174 |
| Coal Transformation | – | – | – | – | – | – | – | – | – | – | – |
| Liquefaction Plants | – | – | 121 | –281 | – | – | – | – | – | – | –160 |
| Other Transformation | – | – | – | – | – | – | – | – | – | – | – |
| Own Use | – | –76 | –9 | – | – | – | – | – | –17 | – | –102 |
| Distribution Losses | – | –68 | – | – | – | – | – | – | –191 | – | –259 |
| **Total Final Consumption** | **–** | **–** | **5140** | **–** | **–** | **–** | **–** | **697** | **678** | **–** | **6516** |
| **INDUSTRY SECTOR** | **–** | **–** | **842** | **–** | **–** | **–** | **–** | **311** | **189** | **–** | **1342** |
| Iron and Steel | – | – | – | – | – | – | – | – | – | – | – |
| Chemical and Petrochemical | – | – | – | – | – | – | – | – | – | – | – |
| of which: Feedstocks | – | – | – | – | – | – | – | – | – | – | – |
| Non–Ferrous Metals | – | – | – | – | – | – | – | – | – | – | – |
| Non–Metallic Minerals | – | – | – | – | – | – | – | – | – | – | – |
| Transport Equipment | – | – | – | – | – | – | – | – | – | – | – |
| Machinery | – | – | – | – | – | – | – | – | – | – | – |
| Mining and Quarrying | – | – | – | – | – | – | – | – | – | – | – |
| Food and Tobacco | – | – | – | – | – | – | – | – | – | – | – |
| Paper Pulp and Printing | – | – | – | – | – | – | – | – | – | – | – |
| Wood and Wood Products | – | – | – | – | – | – | – | – | – | – | – |
| Construction | – | – | – | – | – | – | – | – | – | – | – |
| Textile and Leather | – | – | – | – | – | – | – | – | – | – | – |
| Non–specified | – | – | 842 | – | – | – | – | 311 | 189 | – | 1342 |
| **TRANSPORT SECTOR** | **–** | **–** | **3279** | **–** | **–** | **–** | **–** | **–** | **–** | **–** | **3279** |
| Air | – | – | 227 | – | – | – | – | – | – | – | 227 |
| Road | – | – | 2753 | – | – | – | – | – | – | – | 2753 |
| Rail | – | – | – | – | – | – | – | – | – | – | – |
| Pipeline Transport | – | – | – | – | – | – | – | – | – | – | – |
| Internal Navigation | – | – | 299 | – | – | – | – | – | – | – | 299 |
| Non–specified | – | – | – | – | – | – | – | – | – | – | – |
| **OTHER SECTORS** | **–** | **–** | **790** | **–** | **–** | **–** | **–** | **387** | **490** | **–** | **1667** |
| Agriculture | – | – | – | – | – | – | – | – | – | – | – |
| Comm. and Publ. Services | – | – | 119 | – | – | – | – | – | 170 | – | 289 |
| Residential | – | – | 670 | – | – | – | – | 387 | 240 | – | 1297 |
| Non–specified | – | – | 1 | – | – | – | – | – | 79 | – | 80 |
| **NON–ENERGY USE** | **–** | **–** | **228** | **–** | **–** | **–** | **–** | **–** | **–** | **–** | **228** |
| in Industry/Transf./Energy | – | – | 228 | – | – | – | – | – | – | – | 228 |
| in Transport | – | – | – | – | – | – | – | – | – | – | – |
| in Other Sectors | – | – | – | – | – | – | – | – | – | – | – |
| ***Electr. Generated – GWh*** | **–** | **–** | **3003** | **–** | **–** | **7610** | **–** | **–** | **–** | **–** | **10613** |
| *Electricity Plants* | – | – | 3003 | – | – | 7610 | – | – | – | – | 10613 |
| *CHP Plants* | – | – | – | – | – | – | – | – | – | – | – |

*Source: IEA (2002), Energy Balances of Non-OECD Countries 1999-2000.*

# PARAGUAY

## Energy Balance, 2000

| | | | | | | | | | | | |
|---|---|---|---|---|---|---|---|---|---|---|---|
| | | | | Thousand tonnes of oil equivalent (ktoe) | | | | | | | |
| SUPPLY AND CONSUMPTION | Coal | Crude Oil | Petroleum Products | Gas | Nuclear | Hydro | Geotherm. Solar etc. | Combust. Renew. & Waste | Electricity | Heat | Total |
| Production | – | – | – | – | – | 4599 | – | 2287 | – | – | 6886 |
| Imports | – | 108 | 1049 | – | – | – | – | – | – | – | 1157 |
| Exports | – | – | – | – | – | – | – | – | –4075 | – | –4075 |
| Intl. Marine Bunkers | – | – | – | – | – | – | – | – | – | – | – |
| Stock Changes | – | –4 | –34 | – | – | – | – | – | – | – | –38 |
| **Total Primary Supply** | **–** | **104** | **1015** | **–** | **–** | **4599** | **–** | **2287** | **–4075** | **–** | **3930** |
| Transfers | – | – | – | – | – | – | – | – | – | – | – |
| Statistical Differences | – | – | – | – | – | – | – | 1 | – | – | 1 |
| Electricity Plants | – | – | –6 | – | – | –4599 | – | –44 | 4603 | – | –46 |
| CHP Plants | – | – | – | – | – | – | – | – | – | – | – |
| Heat Plants | – | – | – | – | – | – | – | – | – | – | – |
| Gas Works | – | – | – | – | – | – | – | – | – | – | – |
| Petroleum Refineries | – | –104 | 103 | – | – | – | – | – | – | – | –1 |
| Coal Transformation | – | – | – | – | – | – | – | – | – | – | – |
| Liquefaction Plants | – | – | – | – | – | – | – | – | – | – | – |
| Other Transformation | – | – | – | – | – | – | – | –47 | – | – | –47 |
| Own Use | – | – | – | – | – | – | – | – | –21 | – | –21 |
| Distribution Losses | – | – | – | – | – | – | – | – | –111 | – | –111 |
| **Total Final Consumption** | **–** | **–** | **1112** | **–** | **–** | **–** | **–** | **2197** | **396** | **–** | **3705** |
| **INDUSTRY SECTOR** | **–** | **–** | **79** | **–** | **–** | **–** | **–** | **1151** | **147** | **–** | **1378** |
| Iron and Steel | – | – | – | – | – | – | – | – | – | – | – |
| Chemical and Petrochemical | – | – | – | – | – | – | – | – | – | – | – |
| of which: Feedstocks | – | – | – | – | – | – | – | – | – | – | – |
| Non–Ferrous Metals | – | – | – | – | – | – | – | – | – | – | – |
| Non–Metallic Minerals | – | – | – | – | – | – | – | – | – | – | – |
| Transport Equipment | – | – | – | – | – | – | – | – | – | – | – |
| Machinery | – | – | – | – | – | – | – | – | – | – | – |
| Mining and Quarrying | – | – | – | – | – | – | – | – | – | – | – |
| Food and Tobacco | – | – | – | – | – | – | – | – | – | – | – |
| Paper Pulp and Printing | – | – | – | – | – | – | – | – | – | – | – |
| Wood and Wood Products | – | – | – | – | – | – | – | – | – | – | – |
| Construction | – | – | – | – | – | – | – | – | – | – | – |
| Textile and Leather | – | – | – | – | – | – | – | – | – | – | – |
| Non–specified | – | – | 79 | – | – | – | – | 1151 | 147 | – | 1378 |
| **TRANSPORT SECTOR** | **–** | **–** | **941** | **–** | **–** | **–** | **–** | **14** | **–** | **–** | **955** |
| Air | – | – | 13 | – | – | – | – | – | – | – | 13 |
| Road | – | – | 916 | – | – | – | – | 7 | – | – | 923 |
| Rail | – | – | – | – | – | – | – | – | – | – | – |
| Pipeline Transport | – | – | – | – | – | – | – | – | – | – | – |
| Internal Navigation | – | – | – | – | – | – | – | – | – | – | – |
| Non–specified | – | – | 12 | – | – | – | – | 7 | – | – | 19 |
| **OTHER SECTORS** | **–** | **–** | **85** | **–** | **–** | **–** | **–** | **1031** | **249** | **–** | **1366** |
| Agriculture | – | – | – | – | – | – | – | – | – | – | – |
| Comm. and Publ. Services | – | – | – | – | – | – | – | 4 | 74 | – | 77 |
| Residential | – | – | 85 | – | – | – | – | 1028 | 175 | – | 1288 |
| Non–specified | – | – | – | – | – | – | – | – | – | – | – |
| **NON–ENERGY USE** | **–** | **–** | **7** | **–** | **–** | **–** | **–** | **–** | **–** | **–** | **7** |
| in Industry/Transf./Energy | – | – | 7 | – | – | – | – | – | – | – | 7 |
| in Transport | – | – | – | – | – | – | – | – | – | – | – |
| in Other Sectors | – | – | – | – | – | – | – | – | – | – | – |
| **Electr. Generated – GWh** | **–** | **–** | **13** | **–** | **–** | **53473** | **–** | **35** | **–** | **–** | **53521** |
| Electricity Plants | – | – | 13 | – | – | 53473 | – | 35 | – | – | 53521 |
| CHP Plants | – | – | – | – | – | – | – | – | – | – | – |

Source: IEA (2002) Energy Balances of Non–OECD Countries 1999–2000.

# PERU

## Energy Balance, 2000

| SUPPLY AND CONSUMPTION | Coal | Crude Oil | Petroleum Products | Gas | Nuclear | Hydro | Geotherm. Solar etc. | Combust. Renew. & Waste | Electricity | Heat | Total |
|---|---|---|---|---|---|---|---|---|---|---|---|
| | | | | Thousand tonnes of oil equivalent (ktoe) | | | | | | | |
| Production | 12 | 5177 | – | 610 | – | 1390 | 53 | 2234 | – | – | 9477 |
| Imports | 588 | 3089 | 1723 | – | – | – | – | – | – | – | 5400 |
| Exports | – | –687 | –1365 | – | – | – | – | – | – | – | –2052 |
| Intl. Marine Bunkers | – | – | –43 | – | – | – | – | – | – | – | –43 |
| Stock Changes | 31 | 36 | –153 | – | – | – | – | – | – | – | –87 |
| **Total Primary Supply** | **630** | **7616** | **161** | **610** | **–** | **1390** | **53** | **2234** | **–** | **–** | **12695** |
| Transfers | – | –174 | 185 | – | – | – | – | – | – | – | 11 |
| Statistical Differences | 17 | 298 | –4 | 12 | – | – | – | – | – | – | 323 |
| Electricity Plants | –99 | – | –686 | –203 | – | –1390 | – | –54 | 1713 | – | –721 |
| CHP Plants | – | – | – | – | – | – | – | – | – | – | – |
| Heat Plants | – | – | – | – | – | – | – | – | – | – | – |
| Gas Works | – | – | – | –166 | – | – | – | – | – | – | –166 |
| Petroleum Refineries | – | –7740 | 7579 | – | – | – | – | – | – | – | –161 |
| Coal Transformation | –100 | – | – | – | – | – | – | – | – | – | –100 |
| Liquefaction Plants | – | – | – | – | – | – | – | – | – | – | – |
| Other Transformation | – | – | – | – | – | – | – | –84 | – | – | –84 |
| Own Use | – | – | –366 | –252 | – | – | – | – | –26 | – | –644 |
| Distribution Losses | – | – | – | – | – | – | – | – | –197 | – | –197 |
| **Total Final Consumption** | **448** | **–** | **6868** | **1** | **–** | **–** | **53** | **2096** | **1490** | **–** | **10956** |
| **INDUSTRY SECTOR** | **440** | **–** | **1742** | **–** | **–** | **–** | **–** | **7** | **846** | **–** | **3035** |
| Iron and Steel | 179 | – | 640 | – | – | – | – | – | 371 | – | 1190 |
| Chemical and Petrochemical | – | – | – | – | – | – | – | – | – | – | – |
| of which: Feedstocks | – | – | – | – | – | – | – | – | – | – | – |
| Non–Ferrous Metals | – | – | – | – | – | – | – | – | – | – | – |
| Non–Metallic Minerals | – | – | 6 | – | – | – | – | – | – | – | 6 |
| Transport Equipment | – | – | – | – | – | – | – | – | – | – | – |
| Machinery | – | – | – | – | – | – | – | – | – | – | – |
| Mining and Quarrying | – | – | – | – | – | – | – | – | – | – | – |
| Food and Tobacco | – | – | – | – | – | – | – | – | – | – | – |
| Paper Pulp and Printing | – | – | – | – | – | – | – | – | – | – | – |
| Wood and Wood Products | – | – | – | – | – | – | – | – | – | – | – |
| Construction | – | – | – | – | – | – | – | – | – | – | – |
| Textile and Leather | – | – | – | – | – | – | – | – | – | – | – |
| Non–specified | 262 | – | 1096 | – | – | – | – | 7 | 475 | – | 1840 |
| **TRANSPORT SECTOR** | **–** | **–** | **3388** | **–** | **–** | **–** | **–** | **–** | **–** | **–** | **3388** |
| Air | – | – | 359 | – | – | – | – | – | – | – | 359 |
| Road | – | – | 2969 | – | – | – | – | – | – | – | 2969 |
| Rail | – | – | – | – | – | – | – | – | – | – | – |
| Pipeline Transport | – | – | – | – | – | – | – | – | – | – | – |
| Internal Navigation | – | – | 60 | – | – | – | – | – | – | – | 60 |
| Non–specified | – | – | – | – | – | – | – | – | – | – | – |
| **OTHER SECTORS** | **8** | **–** | **1617** | **1** | **–** | **–** | **53** | **2089** | **644** | **–** | **4411** |
| Agriculture | 8 | – | 345 | – | – | – | – | 147 | 81 | – | 580 |
| Comm. and Publ. Services | – | – | 246 | – | – | – | – | – | 25 | – | 271 |
| Residential | – | – | 1026 | 1 | – | – | 53 | 1942 | 538 | – | 3560 |
| Non–specified | – | – | – | – | – | – | – | – | – | – | – |
| **NON–ENERGY USE** | **–** | **–** | **122** | **–** | **–** | **–** | **–** | **–** | **–** | **–** | **122** |
| in Industry/Transf./Energy | – | – | 122 | – | – | – | – | – | – | – | 122 |
| in Transport | – | – | – | – | – | – | – | – | – | – | – |
| in Other Sectors | – | – | – | – | – | – | – | – | – | – | – |
| **Electr. Generated – GWh** | **208** | **–** | **2670** | **709** | **–** | **16168** | **1** | **158** | **–** | **–** | **19914** |
| Electricity Plants | 208 | – | 2670 | 709 | – | 16168 | 1 | 158 | – | – | 19914 |
| CHP Plants | – | – | – | – | – | – | – | – | – | – | – |

Source: IEA (2002), Energy Balances of Non-OECD Countries 1999-2000.

# TRINIDAD AND TOBAGO

## Energy Balance, 2000

| | | | | | | Thousand tonnes of oil equivalent (ktoe) | | | | | |
|---|---|---|---|---|---|---|---|---|---|---|---|
| SUPPLY AND CONSUMPTION | Coal | Crude Oil | Petroleum Products | Gas | Nuclear | Hydro | Geotherm. Solar etc. | Combust. Renew. & Waste | Electricity | Heat | Total |
| Production | – | 6834 | – | 11016 | – | – | – | 34 | – | – | 17884 |
| Imports | – | 5016 | 11 | – | – | – | – | – | – | – | 5026 |
| Exports | – | –4682 | –6645 | –2780 | – | – | – | – | – | – | –14107 |
| Intl. Marine Bunkers | – | – | –12 | – | – | – | – | – | – | – | –12 |
| Stock Changes | – | –154 | 28 | – | – | – | – | – | – | – | –127 |
| **Total Primary Supply** | **–** | **7014** | **–6619** | **8237** | **–** | **–** | **–** | **34** | **–** | **–** | **8665** |
| Transfers | – | –134 | 151 | – | – | – | – | – | – | – | 17 |
| Statistical Differences | – | 45 | –118 | – | – | – | – | – | –4 | – | –77 |
| Electricity Plants | – | – | –1 | –1613 | – | – | – | –34 | 473 | – | –1174 |
| CHP Plants | – | – | – | – | – | – | – | – | – | – | – |
| Heat Plants | – | – | – | – | – | – | – | – | – | – | – |
| Gas Works | – | – | – | – | – | – | – | – | – | – | – |
| Petroleum Refineries | – | –8402 | 7670 | – | – | – | – | – | – | – | –732 |
| Coal Transformation | – | – | – | – | – | – | – | – | – | – | – |
| Liquefaction Plants | – | 1488 | – | –2062 | – | – | – | – | – | – | –574 |
| Other Transformation | – | – | – | – | – | – | – | – | – | – | – |
| Own Use | – | – | –375 | –706 | – | – | – | – | –22 | – | –1103 |
| Distribution Losses | – | – | – | – | – | – | – | – | –34 | – | –34 |
| **Total Final Consumption** | **–** | **11** | **708** | **3855** | **–** | **–** | **–** | **–** | **413** | **–** | **4987** |
| **INDUSTRY SECTOR** | **–** | **11** | **71** | **3855** | **–** | **–** | **–** | **–** | **262** | **–** | **4198** |
| Iron and Steel | – | – | – | 507 | – | – | – | – | – | – | 507 |
| Chemical and Petrochemical | – | 11 | – | 3123 | – | – | – | – | – | – | 3134 |
| of which: Feedstocks | – | 11 | – | 3123 | – | – | – | – | – | – | 3134 |
| Non–Ferrous Metals | – | – | – | – | – | – | – | – | – | – | – |
| Non–Metallic Minerals | – | – | – | 107 | – | – | – | – | – | – | 107 |
| Transport Equipment | – | – | – | – | – | – | – | – | – | – | – |
| Machinery | – | – | – | – | – | – | – | – | – | – | – |
| Mining and Quarrying | – | – | – | – | – | – | – | – | – | – | – |
| Food and Tobacco | – | – | – | – | – | – | – | – | – | – | – |
| Paper Pulp and Printing | – | – | – | – | – | – | – | – | – | – | – |
| Wood and Wood Products | – | – | – | – | – | – | – | – | – | – | – |
| Construction | – | – | – | – | – | – | – | – | – | – | – |
| Textile and Leather | – | – | – | – | – | – | – | – | – | – | – |
| Non–specified | – | – | 71 | 118 | – | – | – | – | 262 | – | 451 |
| **TRANSPORT SECTOR** | **–** | **–** | **561** | **–** | **–** | **–** | **–** | **–** | **–** | **–** | **561** |
| Air | – | – | 11 | – | – | – | – | – | – | – | 11 |
| Road | – | – | 551 | – | – | – | – | – | – | – | 551 |
| Rail | – | – | – | – | – | – | – | – | – | – | – |
| Pipeline Transport | – | – | – | – | – | – | – | – | – | – | – |
| Internal Navigation | – | – | – | – | – | – | – | – | – | – | – |
| Non–specified | – | – | – | – | – | – | – | – | – | – | – |
| **OTHER SECTORS** | **–** | **–** | **66** | **–** | **–** | **–** | **–** | **–** | **151** | **–** | **217** |
| Agriculture | – | – | – | – | – | – | – | – | – | – | – |
| Comm. and Publ. Services | – | – | – | – | – | – | – | – | – | – | – |
| Residential | – | – | 66 | – | – | – | – | – | 151 | – | 217 |
| Non–specified | – | – | – | – | – | – | – | – | – | – | – |
| **NON–ENERGY USE** | **–** | **–** | **10** | **–** | **–** | **–** | **–** | **–** | **–** | **–** | **10** |
| in Industry/Transf./Energy | – | – | 10 | – | – | – | – | – | – | – | 10 |
| in Transport | – | – | – | – | – | – | – | – | – | – | – |
| in Other Sectors | – | – | – | – | – | – | – | – | – | – | – |
| **Electr. Generated – GWh** | **–** | **–** | **–** | **5490** | **–** | **–** | **–** | **15** | **–** | **–** | **5505** |
| *Electricity Plants* | – | – | – | 5490 | – | – | – | 15 | – | – | 5505 |
| *CHP Plants* | – | – | – | – | – | – | – | – | – | – | – |

*Source: IEA (2002), Energy Balances of Non–OECD Countries 1999–2000.*

# URUGUAY

## Energy Balance, 2000

| SUPPLY AND CONSUMPTION | Coal | Crude Oil | Petroleum Products | Gas | Nuclear | Hydro | Geotherm. Solar etc. | Combust. Renew. & Waste | Electricity | Heat | Total |
|---|---|---|---|---|---|---|---|---|---|---|---|
| | | | Thousand tonnes of oil equivalent (ktoe) | | | | | | | | |
| Production | – | – | – | – | – | 606 | – | 422 | – | – | 1028 |
| Imports | 1 | 2056 | 284 | 30 | – | – | – | 1 | 114 | – | 2486 |
| Exports | – | – | –144 | – | – | – | – | – | –81 | – | –225 |
| Intl. Marine Bunkers | – | – | –284 | – | – | – | – | – | – | – | –284 |
| Stock Changes | – | 113 | –41 | – | – | – | – | – | – | – | 73 |
| **Total Primary Supply** | **1** | **2169** | **–184** | **30** | **–** | **606** | **–** | **423** | **33** | **–** | **3079** |
| Transfers | – | – | – | – | – | – | – | – | – | – | – |
| Statistical Differences | – | –5 | 1 | – | – | – | – | 1 | –4 | – | –7 |
| Electricity Plants | – | – | –135 | – | – | –606 | – | –7 | 653 | – | –97 |
| CHP Plants | – | – | – | – | – | – | – | – | – | – | – |
| Heat Plants | – | – | – | – | – | – | – | – | – | – | – |
| Gas Works | – | – | –18 | 11 | – | – | – | – | – | – | –7 |
| Petroleum Refineries | – | –2164 | 1935 | – | – | – | – | – | – | – | –229 |
| Coal Transformation | – | – | – | – | – | – | – | – | – | – | – |
| Liquefaction Plants | – | – | – | – | – | – | – | – | – | – | – |
| Other Transformation | – | – | – | – | – | – | – | –1 | – | – | –1 |
| Own Use | – | – | –96 | – | – | – | – | – | –9 | – | –105 |
| Distribution Losses | – | – | –7 | –1 | – | – | – | – | –121 | – | –129 |
| **Total Final Consumption** | **1** | **–** | **1495** | **40** | **–** | **–** | **–** | **415** | **552** | **–** | **2503** |
| **INDUSTRY SECTOR** | **1** | **–** | **200** | **30** | **–** | **–** | **–** | **109** | **137** | **–** | **477** |
| Iron and Steel | – | – | – | – | – | – | – | – | – | – | – |
| Chemical and Petrochemical | – | – | – | – | – | – | – | – | – | – | – |
| of which: Feedstocks | – | – | – | – | – | – | – | – | – | – | – |
| Non–Ferrous Metals | – | – | – | – | – | – | – | – | – | – | – |
| Non–Metallic Minerals | – | – | – | – | – | – | – | – | – | – | – |
| Transport Equipment | – | – | – | – | – | – | – | – | – | – | – |
| Machinery | – | – | – | – | – | – | – | – | – | – | – |
| Mining and Quarrying | – | – | – | – | – | – | – | – | – | – | – |
| Food and Tobacco | – | – | – | – | – | – | – | – | – | – | – |
| Paper Pulp and Printing | – | – | – | – | – | – | – | – | – | – | – |
| Wood and Wood Products | – | – | – | – | – | – | – | – | – | – | – |
| Construction | – | – | – | – | – | – | – | – | – | – | – |
| Textile and Leather | – | – | – | – | – | – | – | – | – | – | – |
| Non–specified | 1 | – | 200 | 30 | – | – | – | 109 | 137 | – | 477 |
| **TRANSPORT SECTOR** | **–** | **–** | **832** | **–** | **–** | **–** | **–** | **–** | **–** | **–** | **832** |
| Air | – | – | 9 | – | – | – | – | – | – | – | 9 |
| Road | – | – | 824 | – | – | – | – | – | – | – | 824 |
| Rail | – | – | – | – | – | – | – | – | – | – | – |
| Pipeline Transport | – | – | – | – | – | – | – | – | – | – | – |
| Internal Navigation | – | – | – | – | – | – | – | – | – | – | – |
| Non–specified | – | – | – | – | – | – | – | – | – | – | – |
| **OTHER SECTORS** | **–** | **–** | **396** | **9** | **–** | **–** | **–** | **306** | **415** | **–** | **1127** |
| Agriculture | – | – | 183 | – | – | – | – | – | 17 | – | 200 |
| Comm. and Publ. Services | – | – | 43 | 4 | – | – | – | 3 | 149 | – | 199 |
| Residential | – | – | 166 | 6 | – | – | – | 303 | 249 | – | 724 |
| Non–specified | – | – | 3 | – | – | – | – | – | – | – | 3 |
| **NON–ENERGY USE** | **–** | **–** | **67** | **–** | **–** | **–** | **–** | **–** | **–** | **–** | **67** |
| in Industry/Transf./Energy | – | – | 67 | – | – | – | – | – | – | – | 67 |
| in Transport | – | – | – | – | – | – | – | – | – | – | – |
| in Other Sectors | – | – | – | – | – | – | – | – | – | – | – |
| ***Electr. Generated – GWh*** | **–** | **–** | **508** | **–** | **–** | **7052** | **–** | **30** | **–** | **–** | **7590** |
| *Electricity Plants* | – | – | 508 | – | – | 7052 | – | 30 | – | – | 7590 |
| *CHP Plants* | – | – | – | – | – | – | – | – | – | – | – |

Source: IEA (2002), Energy Balances of Non-OECD Countries 1999-2000.

# VENEZUELA

## Energy Balance, 2000

Thousand tonnes of oil equivalent (ktoe)

| SUPPLY AND CONSUMPTION | Coal | Crude Oil | Petroleum Products | Gas | Nuclear | Hydro | Geotherm. Solar etc. | Combust. Renew. & Waste | Electricity | Heat | Total |
|---|---|---|---|---|---|---|---|---|---|---|---|
| Production | 5756 | 186701 | – | 27069 | – | 5403 | – | 541 | – | – | 225470 |
| Imports | – | – | – | – | – | – | – | – | – | – | – |
| Exports | –5789 | –119653 | –39717 | – | – | – | – | – | – | – | –165159 |
| Intl. Marine Bunkers | – | – | –645 | – | – | – | – | – | – | – | –645 |
| Stock Changes | 164 | –721 | 147 | – | – | – | – | – | – | – | –410 |
| **Total Primary Supply** | **131** | **66328** | **–40215** | **27069** | **–** | **5403** | **–** | **541** | **–** | **–** | **59256** |
| Transfers | – | –6994 | 6438 | – | – | – | – | – | – | – | –556 |
| Statistical Differences | 1 | –1 | 471 | – | – | – | – | – | – | – | 471 |
| Electricity Plants | – | – | –2185 | –4744 | – | –5403 | – | – | 7328 | – | –5004 |
| CHP Plants | – | – | – | – | – | – | – | – | – | – | – |
| Heat Plants | – | – | – | – | – | – | – | – | – | – | – |
| Gas Works | – | – | – | – | – | – | – | – | – | – | – |
| Petroleum Refineries | – | –59331 | 57418 | – | – | – | – | – | – | – | –1913 |
| Coal Transformation | – | – | – | – | – | – | – | – | – | – | – |
| Liquefaction Plants | – | – | – | – | – | – | – | – | – | – | – |
| Other Transformation | – | – | – | – | – | – | – | –15 | – | – | –15 |
| Own Use | – | – | –3356 | –11340 | – | – | – | – | –283 | – | –14980 |
| Distribution Losses | – | –2 | – | – | – | – | – | – | –1779 | – | –1781 |
| **Total Final Consumption** | **132** | **–** | **18571** | **10984** | **–** | **–** | **–** | **525** | **5265** | **–** | **35478** |
| **INDUSTRY SECTOR** | **132** | **–** | **3199** | **9575** | **–** | **–** | **–** | **331** | **2410** | **–** | **15647** |
| Iron and Steel | – | – | – | 3071 | – | – | – | – | 578 | – | 3649 |
| Chemical and Petrochemical | – | – | 1253 | 3308 | – | – | – | – | 52 | – | 4613 |
| of which: Feedstocks | – | – | – | – | – | – | – | – | – | – | – |
| Non-Ferrous Metals | – | – | 96 | 512 | – | – | – | – | 860 | – | 1468 |
| Non-Metallic Minerals | 132 | – | 80 | 874 | – | – | – | – | – | – | 1086 |
| Transport Equipment | – | – | – | – | – | – | – | – | – | – | – |
| Machinery | – | – | – | – | – | – | – | – | – | – | – |
| Mining and Quarrying | – | – | – | – | – | – | – | – | – | – | – |
| Food and Tobacco | – | – | – | – | – | – | – | 331 | – | – | 331 |
| Paper Pulp and Printing | – | – | – | – | – | – | – | – | – | – | – |
| Wood and Wood Products | – | – | – | – | – | – | – | – | – | – | – |
| Construction | – | – | – | – | – | – | – | – | – | – | – |
| Textile and Leather | – | – | – | – | – | – | – | – | – | – | – |
| Non-specified | – | – | 1770 | 1809 | – | – | – | – | 921 | – | 4500 |
| **TRANSPORT SECTOR** | **–** | **–** | **11889** | **98** | **–** | **–** | **–** | **–** | **22** | **–** | **12008** |
| Air | – | – | 316 | – | – | – | – | – | – | – | 316 |
| Road | – | – | 11561 | – | – | – | – | – | – | – | 11561 |
| Rail | – | – | 11 | – | – | – | – | – | 22 | – | 33 |
| Pipeline Transport | – | – | – | – | – | – | – | – | – | – | – |
| Internal Navigation | – | – | – | – | – | – | – | – | – | – | – |
| Non-specified | – | – | – | 98 | – | – | – | – | – | – | 98 |
| **OTHER SECTORS** | **–** | **–** | **1260** | **1312** | **–** | **–** | **–** | **195** | **2834** | **–** | **5600** |
| Agriculture | – | – | 159 | – | – | – | – | – | – | – | 159 |
| Comm. and Publ. Services | – | – | 31 | 379 | – | – | – | – | 1523 | – | 1932 |
| Residential | – | – | 1070 | 933 | – | – | – | 195 | 1311 | – | 3509 |
| Non-specified | – | – | – | – | – | – | – | – | – | – | – |
| **NON-ENERGY USE** | **–** | **–** | **2224** | **–** | **–** | **–** | **–** | **–** | **–** | **–** | **2224** |
| in Industry/Transf./Energy | – | – | 2224 | – | – | – | – | – | – | – | 2224 |
| in Transport | – | – | – | – | – | – | – | – | – | – | – |
| in Other Sectors | – | – | – | – | – | – | – | – | – | – | – |
| ***Electr. Generated - GWh*** | **–** | **–** | **8350** | **14035** | **–** | **62826** | **–** | **–** | **–** | **–** | **85211** |
| *Electricity Plants* | – | – | 8350 | 14035 | – | 62826 | – | – | – | – | 85211 |
| *CHP Plants* | – | – | – | – | – | – | – | – | – | – | – |

*Source: IEA (2002), Energy Balances of Non-OECD Countries 1999-2000.*

# ANNEX 2
# NATURAL GAS STATISTICS

## SOUTH AMERICA

## Natural Gas Supply and Consumption

| | 1980 | 1990 | 1997 | 1998 | 1999 | 2000 | 2001 |
|---|---|---|---|---|---|---|---|
| | | | Million cubic metres (mcm) | | | | |
| **Indigenous Production** | 35944 | 61126 | 84825 | 91184 | 94897 | 103030 | 106950 |
| + Imports | 2174 | 2175 | 2407 | 3812 | 4636 | 7172 | 8627 |
| – Exports | 2286 | 2569 | 2318 | 3528 | 5685 | 10085 | 11338 |
| + Stock Changes | – 2 | – 8 | – | – | – | – | – |
| – Statistical Difference | 406 | 1233 | 43 | – 332 | 389 | 31 | – 24 |
| **Total Primary Supply** | 35424 | 59491 | 84871 | 91800 | 93459 | 100086 | 104263 |
| **Power Generation** | 8756 | 15523 | 20361 | 21950 | 22279 | 23446 | – |
| **Oil and Gas Industry** | 9773 | 13971 | 21282 | 23773 | 22200 | 25185 | – |
| Oil and Gas Extraction | 4960 | 10852 | 16097 | 18584 | 16218 | 19899 | – |
| Gas Inputs to Oil Refineries | 3047 | 1947 | 2524 | 2337 | 2326 | 2428 | – |
| Pipelines Consumption | – | – | 883 | 929 | 991 | 1064 | – |
| Distribution Losses | 1048 | 359 | 1220 | 1127 | 1663 | 878 | – |
| Other | 718 | 813 | 558 | 796 | 1002 | 916 | – |
| **Final Consumption** | 16894 | 29998 | 43230 | 46078 | 48978 | 51455 | – |
| **Transport** | – | 242 | 1517 | 1784 | 1891 | 2198 | – |
| **Industry** | 13374 | 22713 | 32704 | 35116 | 36204 | 37859 | – |
| Iron and Steel | 1327 | 3863 | 5861 | 6084 | 5503 | 6369 | – |
| Chemical and Petrochem. | 3393 | 5888 | 10024 | 10894 | 11733 | 11424 | – |
| Methanol Production | 35 | 2939 | 5468 | 5876 | 7025 | 7661 | – |
| Other | 8619 | 10023 | 11351 | 12262 | 11943 | 12405 | – |
| **Other Sectors** | 3520 | 7043 | 9009 | 9178 | 10883 | 11398 | – |
| Commerce and Public | 20 | 2263 | 2257 | 2161 | 2199 | 2388 | – |
| Residential | 3500 | 4780 | 6739 | 7017 | 8684 | 9010 | – |
| Other | 5 | 4 | 13 | 5 | 5 | 6 | – |

Note: South America is defined here as including Argentina, Bolivia, Brazil, Chile, Colombia, Ecuador, Paraguay, Peru, Trinidad & Tobago, Uruguay and Venezuela.
Source: IEA Gas Statistics Database.

## ARGENTINA

### Natural Gas Supply and Consumption

| Million cubic metres (mcm) | 1980 | 1990 | 1997 | 1998 | 1999 | 2000 | 2001 |
|---|---|---|---|---|---|---|---|
| **Indigenous Production** | 9858 | 20326 | 30648 | 32373 | 38259 | 40465 | 41503 |
| + Imports | 2174 | 2175 | 1703 | 1805 | 420 | 598 | 600 |
| − Exports | – | – | 669 | 1918 | 3384 | 4642 | 5714 |
| + Stock Changes | – | – | – | – | – | – | – |
| − Statistical Difference | 350 | 2 | 484 | 227 | 95 | − 102 | − 11 |
| **Total Primary Supply** | 11682 | 22499 | 31198 | 32033 | 35200 | 36523 | 36400 |
| **Power Generation** | 2778 | 6243 | 8586 | 8514 | 10665 | 10848 | – |
| **Oil and Gas Industry** | 2668 | 3349 | 4286 | 4810 | 5042 | 5327 | – |
| Oil and Gas Extraction | 1756 | 3074 | 3107 | 3557 | 3738 | 3959 | – |
| Gas Inputs to Oil Refineries | – | – | – | – | – | – | – |
| Pipelines Consumption | – | – | 883 | 929 | 991 | 1064 | – |
| Distribution Losses | 912 | 275 | 296 | 324 | 313 | 304 | – |
| Other | – | – | – | – | – | – | – |
| **Final Consumption** | 6236 | 12906 | 18328 | 18711 | 19493 | 20350 | – |
| **Transport** | – | 216 | 1268 | 1412 | 1509 | 1672 | – |
| **Industry** | 3482 | 6613 | 9742 | 9908 | 9795 | 10021 | – |
| Iron and Steel | – | – | 1651 | 1771 | 1487 | 1666 | – |
| Chemical and Petrochem. | 129 | 177 | 1112 | 1122 | 1122 | 1404 | – |
| Methanol Production | – | 1470 | 2344 | 2109 | 2281 | 2172 | – |
| Other | 3353 | 4966 | 4635 | 4906 | 4905 | 4779 | – |
| **Other Sectors** | 2754 | 6077 | 7318 | 7391 | 8189 | 8657 | – |
| Commerce and Public | – | 1767 | 1500 | 1399 | 1517 | 1710 | – |
| Residential | 2754 | 4310 | 5818 | 5992 | 6672 | 6947 | – |
| Other | – | – | – | – | – | – | – |

## BOLIVIA

### Natural Gas Supply and Consumption

| Million cubic metres (mcm) | 1980 | 1990 | 1997 | 1998 | 1999 | 2000 | 2001 |
|---|---|---|---|---|---|---|---|
| **Indigenous Production** | 2581 | 3333 | 3572 | 3487 | 2860 | 3711 | 4471 |
| + Imports | – | – | – | – | – | – | – |
| − Exports | 2286 | 2569 | 1602 | 1595 | 1042 | 2121 | 2574 |
| + Stock Changes | − 2 | − 8 | – | – | – | – | – |
| − Statistical Difference | − 3 | − 22 | − 3 | 1 | 267 | 146 | 178 |
| **Total Primary Supply** | 296 | 778 | 1973 | 1891 | 1551 | 1444 | 1719 |
| **Power Generation** | 106 | 300 | 702 | 800 | 549 | 554 | – |
| **Oil and Gas Industry** | 145 | 279 | 867 | 656 | 554 | 468 | – |
| Oil and Gas Extraction | 145 | 276 | 114 | 139 | 162 | 193 | – |
| Gas Inputs to Oil Refineries | – | – | 12 | 13 | 15 | 14 | – |
| Pipelines Consumption | – | – | – | – | – | – | – |
| Distribution Losses | – | 3 | 741 | 504 | 377 | 261 | – |
| Other | – | – | – | – | – | – | – |
| **Final Consumption** | 45 | 199 | 403 | 436 | 447 | 420 | – |
| **Transport** | – | – | 30 | 16 | 20 | 26 | – |
| **Industry** | 45 | 199 | 357 | 411 | 416 | 382 | – |
| Iron and Steel | – | – | – | – | – | – | – |
| Chemical and Petrochem. | – | – | – | – | – | – | – |
| Methanol Production | – | – | – | – | – | – | – |
| Other | 45 | 199 | 357 | 411 | 416 | 382 | – |
| **Other Sectors** | – | – | 16 | 9 | 11 | 12 | – |
| Commerce and Public | – | – | 9 | 5 | 6 | 7 | – |
| Residential | – | – | 7 | 4 | 5 | 5 | – |
| Other | – | – | – | – | – | – | – |

*Note: Totals may not add up due to rounding.*
*Source: IEA Gas Statistics Database.*

## BRAZIL
# Natural Gas Supply and Consumption

| Million cubic metres (mcm) | 1980 | 1990 | 1997 | 1998 | 1999 | 2000 | 2001 |
|---|---|---|---|---|---|---|---|
| **Indigenous Production** | 988 | 2885 | 4210 | 5686 | 6109 | 6793 | 7697 |
| + Imports | – | – | – | – | 400 | 2166 | 3022 |
| – Exports | – | – | – | – | – | – | – |
| + Stock Changes | – | – | – | – | – | – | – |
| – Statistical Difference | – 15 | – | 9 | 2 | 31 | – 11 | 20 |
| **Total Primary Supply** | 1003 | 2885 | 4201 | 5684 | 6478 | 8970 | 10699 |
| **Power Generation** | – | 86 | 31 | 50 | 317 | 639 | – |
| **Oil and Gas Industry** | 188 | 532 | 522 | 1785 | 1543 | 2646 | – |
| Oil and Gas Extraction | 188 | 339 | 246 | 1471 | 1275 | 2367 | – |
| Gas Inputs to Oil Refineries | – | – | – | – | – | – | – |
| Pipelines Consumption | – | – | – | – | – | – | – |
| Distribution Losses | – | – | 180 | 159 | 135 | 183 | – |
| Other | – | 193 | 96 | 155 | 133 | 96 | – |
| **Final Consumption** | 815 | 2268 | 3649 | 3850 | 4619 | 5687 | – |
| **Transport** | – | 2 | 47 | 132 | 159 | 313 | – |
| **Industry** | 815 | 2258 | 3429 | 3560 | 4324 | 5174 | – |
| Iron and Steel | 129 | 383 | 804 | 687 | 731 | 832 | – |
| Chemical and Petrochem. | 631 | 1227 | 1330 | 1595 | 2027 | 2254 | – |
| Methanol Production | – | – | – | – | – | – | – |
| Other | 55 | 648 | 1295 | 1278 | 1566 | 2088 | – |
| **Other Sectors** | – | 8 | 173 | 158 | 136 | 200 | – |
| Commerce and Public | – | 3 | 92 | 71 | 57 | 86 | – |
| Residential | – | 5 | 81 | 87 | 79 | 114 | – |
| Other | – | – | – | – | – | – | – |

## CHILE
# Natural Gas Supply and Consumption

| Million cubic metres (mcm) | 1980 | 1990 | 1997 | 1998 | 1999 | 2000 | 2001 |
|---|---|---|---|---|---|---|---|
| **Indigenous Production** | 856 | 1771 | 1922 | 1733 | 1972 | 2085 | 1738 |
| + Imports | – | – | 704 | 2005 | 3792 | 4371 | 4940 |
| - Exports | – | – | 47 | 15 | – | – | – |
| + Stock Changes | – | – | – | – | – | – | – |
| – Statistical Difference | – | – | – | – | – | – | – 199 |
| **Total Primary Supply** | 856 | 1771 | 2579 | 3723 | 5764 | 6456 | 6877 |
| **Power Generation** | 47 | 74 | 197 | 961 | 1357 | 1966 | – |
| **Oil and Gas Industry** | 691 | 624 | 455 | 565 | 1257 | 571 | – |
| Oil and Gas Extraction | 555 | 543 | 452 | 425 | 419 | 441 | – |
| Gas Inputs to Oil Refineries | – | – | – | – | – | – | – |
| Pipelines Consumption | – | – | – | – | – | – | – |
| Distribution Losses | 136 | 81 | 3 | 140 | 838 | 130 | – |
| Other | – | – | – | – | – | – | – |
| **Final Consumption** | 118 | 1073 | 1928 | 2196 | 3151 | 3918 | – |
| **Transport** | – | 7 | 6 | 7 | 7 | 9 | – |
| **Industry** | 9 | 887 | 1690 | 1965 | 2872 | 3553 | – |
| Iron and Steel | – | – | – | 2 | 4 | 9 | – |
| Chemical and Petrochem. | – | – | – | 32 | 43 | 13 | – |
| Methanol Production | – | 884 | 1380 | 1628 | 2461 | 2754 | – |
| Other | 9 | 3 | 310 | 303 | 364 | 777 | – |
| **Other Sectors** | 109 | 179 | 232 | 224 | 272 | 356 | – |
| Commerce and Public | 20 | 33 | 42 | 41 | 50 | 68 | – |
| Residential | 89 | 146 | 190 | 183 | 222 | 288 | – |
| Other | – | – | – | – | – | – | – |

*Note: Totals may not add up due to rounding.*
*Source: IEA Gas Statistics Database.*

## COLOMBIA

# Natural Gas Supply and Consumption

| | Million cubic metres (mcm) | | | | | | |
|---|---|---|---|---|---|---|---|
| | 1980 | 1990 | 1997 | 1998 | 1999 | 2000 | 2001 |
| **Indigenous Production** | 3219 | 4537 | 6463 | 7503 | 6582 | 7338 | 7982 |
| + Imports | – | – | – | – | – | – | – |
| – Exports | – | – | – | – | – | – | – |
| + Stock Changes | – | – | – | – | – | – | – |
| – Statistical Difference | – 61 | 211 | 14 | – 114 | – 5 | – 2 | – 2 |
| **Total Primary Supply** | 3280 | 4326 | 6449 | 7617 | 6587 | 7340 | 7984 |
| **Power Generation** | 1465 | 1672 | 2928 | 3105 | 1923 | 2538 | – |
| **Oil and Gas Industry** | 1086 | 1426 | 1735 | 2576 | 2597 | 2617 | – |
| Oil and Gas Extraction | 326 | 386 | 545 | 1402 | 1322 | 1346 | – |
| Gas Inputs to Oil Refineries | 760 | 1040 | 1190 | 1174 | 1275 | 1271 | – |
| Pipelines Consumption | – | – | – | – | – | – | – |
| Distribution Losses | – | – | – | – | – | – | – |
| Other | – | – | – | – | – | – | – |
| **Final Consumption** | 729 | 1230 | 1786 | 1935 | 2065 | 2185 | – |
| **Transport** | – | 17 | 60 | 62 | 65 | 76 | – |
| **Industry** | 726 | 1069 | 1222 | 1256 | 1265 | 1312 | – |
| Iron and Steel | 12 | 33 | 38 | 35 | 35 | 36 | – |
| Chemical and Petrochem. | 339 | 466 | 527 | 530 | 534 | 553 | – |
| Methanol Production | – | – | – | – | – | – | – |
| Other | 375 | 570 | 657 | 691 | 696 | 723 | – |
| **Other Sectors** | 3 | 144 | 504 | 617 | 735 | 797 | – |
| Commerce and Public | – | 22 | 74 | 93 | 110 | 120 | – |
| Residential | 3 | 122 | 417 | 524 | 625 | 677 | – |
| Other | 5 | 4 | 13 | 5 | 5 | 6 | – |

## ECUADOR

# Natural Gas Supply and Consumption

| | Million cubic metres (mcm) | | | | | | |
|---|---|---|---|---|---|---|---|
| | 1980 | 1990 | 1997 | 1998 | 1999 | 2000 | 2001 |
| **Indigenous Production** | 35 | 216 | 258 | 263 | 247 | 270 | 270 |
| + Imports | – | – | – | – | – | – | – |
| – Exports | – | – | – | – | – | – | – |
| + Stock Changes | – | – | – | – | – | – | – |
| – Statistical Difference | – | – | – | – | – | – | – |
| **Total Primary Supply** | 35 | 216 | 258 | 263 | 247 | 270 | 270 |
| **Power Generation** | – | – | – | – | – | – | – |
| **Oil and Gas Industry** | – | – | – | – | – | – | – |
| Oil and Gas Extraction | – | – | – | – | – | – | – |
| Gas Inputs to Oil Refineries | – | – | – | – | – | – | – |
| Pipelines Consumption | – | – | – | – | – | – | – |
| Distribution Losses | – | – | – | – | – | – | – |
| Other | – | – | – | – | – | – | – |
| **Final Consumption** | 35 | 216 | 258 | 263 | 247 | 270 | – |
| **Transport** | – | – | – | – | – | – | – |
| **Industry** | 35 | 216 | 258 | 263 | 247 | 270 | – |
| Iron and Steel | – | – | – | – | – | – | – |
| Chemical and Petrochem. | – | – | – | – | – | – | – |
| Methanol Production | 35 | 216 | 258 | 263 | 247 | 270 | – |
| Other | – | – | – | – | – | – | – |
| **Other Sectors** | – | – | – | – | – | – | – |
| Commerce and Public | – | – | – | – | – | – | – |
| Residential | – | – | – | – | – | – | – |
| Other | – | – | – | – | – | – | – |

Note: Totals may not add up due to rounding.
Source: IEA Gas Statistics Database.

# PERU

## Natural Gas Supply and Consumption

| | Million cubic metres (mcm) | | | | | | |
|---|---|---|---|---|---|---|---|
| | **1980** | **1990** | **1997** | **1998** | **1999** | **2000** | **2001** |
| **Indigenous Production** | 853 | 691 | 463 | 641 | 869 | 820 | 868 |
| + Imports | – | – | – | – | – | – | – |
| – Exports | – | – | – | – | – | – | – |
| + Stock Changes | – | – | – | – | – | – | – |
| – Statistical Difference | 135 | 71 | 1 | – | – | – | – |
| **Total Primary Supply** | **718** | **620** | **462** | **641** | **869** | **820** | **868** |
| **Power Generation** | – | – | – | – | – | – | – |
| **Oil and Gas Industry** | 718 | 620 | 462 | 641 | 869 | 820 | – |
| Oil and Gas Extraction | – | – | – | – | – | – | – |
| Gas Inputs to Oil Refineries | – | – | – | – | – | – | – |
| Pipelines Consumption | – | – | – | – | – | – | – |
| Distribution Losses | – | – | – | – | – | – | – |
| Non–specified | 718 | 620 | 462 | 641 | 869 | 820 | – |
| **Final Consumption** | – | – | – | – | – | – | – |
| **Transport** | – | – | – | – | – | – | – |
| **Industry** | – | – | – | – | – | – | – |
| Iron and Steel | – | – | – | – | – | – | – |
| Chemical and Petrochem. | – | – | – | – | – | – | – |
| Methanol Production | – | – | – | – | – | – | – |
| Other | – | – | – | – | – | – | – |
| **Other Sectors** | – | – | – | – | – | – | – |
| Commerce and Public | – | – | – | – | – | – | – |
| Residential | – | – | – | – | – | – | – |
| Other | – | – | – | – | – | – | – |

# TRINIDAD AND TOBAGO

## Natural Gas Supply and Consumption

| | Million cubic metres (mcm) | | | | | | |
|---|---|---|---|---|---|---|---|
| | **1980** | **1990** | **1997** | **1998** | **1999** | **2000** | **2001** |
| **Indigenous Production** | 2911 | 5613 | 7210 | 8470 | 10797 | 13166 | 13100 |
| + Imports | – | – | – | – | – | – | – |
| – Exports | – | – | – | – | 1259 | 3322 | 3050 |
| + Stock Changes | – | – | – | – | – | – | – |
| – Statistical Difference | – | – | – 442 | – 458 | 1 | – | – 10 |
| **Total Primary Supply** | **2911** | **5613** | **7652** | **8928** | **9537** | **9844** | **10060** |
| **Power Generation** | 713 | 1295 | 1730 | 1868 | 1894 | 1927 | – |
| **Oil and Gas Industry** | 1167 | 952 | 952 | 956 | 811 | 845 | – |
| Oil and Gas Extraction | 382 | 558 | 412 | 434 | 372 | 357 | – |
| Gas Inputs to Oil Refineries | 785 | 394 | 540 | 522 | 439 | 488 | – |
| Pipelines Consumption | – | – | – | – | – | – | – |
| Distribution Losses | – | – | – | – | – | – | – |
| Other | – | – | – | – | – | – | – |
| **Final Consumption** | **1030** | **3365** | **4969** | **6102** | **6832** | **7072** | – |
| **Transport** | – | – | – | – | – | – | – |
| **Industry** | 1030 | 3365 | 4969 | 6102 | 6832 | 7072 | – |
| Iron and Steel | – | 266 | 500 | 489 | 588 | 606 | – |
| Chemical and Petrochem. | 931 | 2270 | 2735 | 3493 | 3945 | 3732 | – |
| Methanol Production | – | 369 | 1486 | 1876 | 2036 | 2465 | – |
| Other | 99 | 460 | 248 | 244 | 263 | 269 | – |
| **Other Sectors** | – | – | – | – | – | – | – |
| Commerce and Public | – | – | – | – | – | – | – |
| Residential | – | – | – | – | – | – | – |
| Other | – | – | – | – | – | – | – |

Note: Totals may not add up due to rounding.
Source: IEA Gas Statistics Database.

## URUGUAY
## Natural Gas Supply and Consumption

| Million cubic metres (mcm) | | | | | | | |
|---|---|---|---|---|---|---|---|
| | 1980 | 1990 | 1997 | 1998 | 1999 | 2000 | 2001 |
| **Indigenous Production** | – | – | – | – | – | – | – |
| + Imports | – | – | – | 2 | 24 | 37 | 65 |
| – Exports | – | – | – | – | – | – | – |
| + Stock Changes | – | – | – | – | – | – | – |
| – Statistical Difference | – | – | – | – | – | – | – |
| **Total Primary Supply** | – | – | – | 2 | 24 | 37 | 65 |
| **Power Generation** | – | – | – | – | – | – | – |
| **Oil and Gas Industry** | – | – | – | – | – | – | – |
| Oil and Gas Extraction | – | – | – | – | – | – | – |
| Gas Inputs to Oil Refineries | – | – | – | – | – | – | – |
| Pipelines Consumption | – | – | – | – | – | – | – |
| Distribution Losses | – | – | – | – | – | – | – |
| Other | – | – | – | – | – | – | – |
| **Final Consumption** | – | – | – | 2 | 23 | 36 | – |
| **Transport** | – | – | – | – | – | – | – |
| **Industry** | – | – | – | 2 | 23 | 36 | – |
| Iron and Steel | – | – | – | – | – | – | – |
| Chemical and Petrochem. | – | – | – | – | – | – | – |
| Methanol Production | – | – | – | – | – | – | – |
| Other | – | – | – | 2 | 23 | 36 | – |
| **Other Sectors** | – | – | – | – | – | – | – |
| Commerce and Public | – | – | – | – | – | – | – |
| Residential | – | – | – | – | – | – | – |
| Other | – | – | – | – | – | – | – |

## VENEZUELA
## Natural Gas Supply and Consumption

| Million cubic metres (mcm) | | | | | | | |
|---|---|---|---|---|---|---|---|
| | 1980 | 1990 | 1997 | 1998 | 1999 | 2000 | 2001 |
| **Indigenous Production** | 14643 | 21754 | 30079 | 31028 | 27202 | 28382 | 29321 |
| + Imports | – | – | – | – | – | – | – |
| – Exports | – | – | – | – | – | – | – |
| + Stock Changes | – | – | – | – | – | – | – |
| – Statistical Difference | – | 971 | – 20 | 10 | – | – | – |
| **Total Primary Supply** | 14643 | 20783 | 30099 | 31018 | 27202 | 28382 | 29321 |
| **Power Generation** | 3647 | 5853 | 6187 | 6652 | 5574 | 4974 | – |
| **Oil and Gas Industry** | 3110 | 6189 | 12003 | 11784 | 9527 | 11891 | – |
| Oil and Gas Extraction | 1608 | 5676 | 11221 | 11156 | 8930 | 11236 | – |
| Gas Inputs to Oil Refineries | 1502 | 513 | 782 | 628 | 597 | 655 | – |
| Pipelines Consumption | – | – | – | – | – | – | – |
| Distribution Losses | – | – | – | – | – | – | – |
| Other | – | – | – | – | – | – | – |
| **Final Consumption** | 7886 | 8741 | 11909 | 12583 | 12101 | 11517 | – |
| **Transport** | – | – | 106 | 155 | 131 | 102 | – |
| **Industry** | 7232 | 8106 | 11037 | 11649 | 10430 | 10039 | – |
| Iron and Steel | 1186 | 3181 | 2868 | 3100 | 2658 | 3220 | – |
| Chemical and Petrochem. | 1363 | 1748 | 4320 | 4122 | 4062 | 3468 | – |
| Methanol Production | – | – | – | – | – | – | – |
| Other | 4683 | 3177 | 3849 | 4427 | 3710 | 3351 | – |
| **Other Sectors** | 654 | 635 | 766 | 779 | 1540 | 1376 | – |
| Commerce and Public | – | 438 | 540 | 552 | 459 | 397 | – |
| Residential | 654 | 197 | 226 | 227 | 1081 | 979 | – |
| Other | – | – | – | – | – | – | – |

Note: Totals may not add up due to rounding.
Source: IEA Gas Statistics Database.

# ANNEX 3
# ENERGY INDICATORS

| | Population (millions) | | | GDP billion 1995 US$ using exchange rates | | | GDP billion 1995 US$ using PPPs | | | Per Capita GDP thousand 1995 US$ using ex. rates | | Per Capita GDP thousand 1995 US$ using PPPs | |
|---|---|---|---|---|---|---|---|---|---|---|---|---|---|
| | 1990 | 2000 | a.a.g.r. 1990-2000 | 1990 | 2000 | a.a.g.r. 1990-2000 | 1990 | 2000 | a.a.g.r. 1990-2000 | 1990 | 2000 | 1990 | 2000 |
| **SOUTH AMERICA** | | | | | | | | | | | | | |
| Argentina | 32.5 | 37.0 | 1.3% | 188 | 294 | 4.6% | 272 | 426 | 4.6% | 5.8 | 7.9 | 8.4 | 11.5 |
| Bolivia | 6.6 | 8.3 | 2.4% | 5 | 8 | 3.8% | 13 | 19 | 3.8% | 0.8 | 1.0 | 2.0 | 2.3 |
| Brazil | 148.0 | 170.4 | 1.4% | 603 | 788 | 2.7% | 974 | 1184 | 2.0% | 4.1 | 4.6 | 6.6 | 6.9 |
| Chile | 13.1 | 15.2 | 1.5% | 43 | 81 | 6.6% | 71 | 135 | 6.6% | 3.3 | 5.4 | 5.5 | 8.9 |
| Colombia | 35.0 | 42.3 | 1.9% | 74 | 97 | 2.7% | 189 | 247 | 2.7% | 2.1 | 2.3 | 5.4 | 5.8 |
| Ecuador | 10.3 | 12.6 | 2.1% | 15 | 18 | 1.8% | 31 | 36 | 1.8% | 1.5 | 1.4 | 3.0 | 2.9 |
| Paraguay | 4.2 | 5.5 | 2.7% | 8 | 9 | 2.0% | 19 | 23 | 2.0% | 1.8 | 1.7 | 4.4 | 4.1 |
| Peru | 21.6 | 25.7 | 1.8% | 41 | 61 | 4.0% | 78 | 116 | 4.0% | 1.9 | 2.4 | 3.6 | 4.5 |
| Trinidad & Tobago | 1.2 | 1.3 | 0.7% | 5 | 7 | 3.0% | 8 | 11 | 3.0% | 4.1 | 5.1 | 6.7 | 8.4 |
| Uruguay | 3.1 | 3.3 | 0.7% | 15 | 20 | 3.0% | 21 | 28 | 3.0% | 4.9 | 6.1 | 6.7 | 8.5 |
| Venezuela | 19.5 | 24.2 | 2.2% | 65 | 80 | 2.0% | 109 | 133 | 2.0% | 3.3 | 3.3 | 5.6 | 5.5 |
| **Total South America** | **295.0** | **345.9** | **1.6%** | **1063** | **1463** | **3.2%** | **1786** | **2359** | **2.8%** | **3.6** | **4.2** | **6.1** | **6.8** |
| **For comparison:** | | | | | | | | | | | | | |
| **IEA COUNTRIES** | | | | | | | | | | | | | |
| Canada | 27.7 | 30.8 | 1.0% | 536 | 705 | 2.8% | 622 | 818 | 2.8% | 19.4 | 22.9 | 22.5 | 26.6 |
| Netherlands | 14.9 | 15.9 | 0.6% | 373 | 497 | 2.9% | 296 | 394 | 2.9% | 25.0 | 31.2 | 19.8 | 24.7 |
| Norway | 4.2 | 4.5 | 0.6% | 122 | 170 | 3.4% | 85 | 118 | 3.4% | 28.8 | 38.0 | 20.0 | 26.3 |
| UK | 57.6 | 59.8 | 0.4% | 1040 | 1304 | 2.3% | 1008 | 1263 | 2.3% | 18.1 | 21.8 | 17.5 | 21.1 |
| US | 250.0 | 275.4 | 1.0% | 6521 | 8987 | 3.3% | 6521 | 8987 | 3.3% | 26.1 | 32.6 | 26.1 | 32.6 |
| IEA North America | 277.7 | 306.2 | 1.0% | 7057 | 9692 | 3.2% | 7143 | 9805 | 3.2% | 25.4 | 31.7 | 25.7 | 32.0 |
| IEA Europe | 453.7 | 477.3 | 0.5% | 8691 | 10606 | 2.0% | 7554 | 9254 | 2.0% | 19.2 | 22.2 | 16.7 | 19.4 |
| IEA Asia-Pacific | 186.9 | 197.2 | 0.5% | 5648 | 6818 | 1.9% | 3480 | 4341 | 2.2% | 30.2 | 34.6 | 18.6 | 22.0 |
| **IEA Total** | **918.2** | **980.6** | **0.7%** | **21395** | **27116** | **2.4%** | **18177** | **23399** | **2.6%** | **23.3** | **27.7** | **19.8** | **23.9** |
| **OTHER COUNTRIES** | | | | | | | | | | | | | |
| China | 1135.2 | 1262.5 | 1.1% | 396 | 1040 | 10.1% | 1799 | 4721 | 10.1% | 0.3 | 0.8 | 1.6 | 3.7 |
| India | 849.5 | 1015.9 | 1.8% | 274 | 467 | 5.5% | 1321 | 2247 | 5.5% | 0.3 | 0.5 | 1.6 | 2.2 |
| Indonesia | 178.2 | 210.4 | 1.7% | 138 | 209 | 4.2% | 381 | 576 | 4.2% | 0.8 | 1.0 | 2.1 | 2.7 |
| Mexico | 81.7 | 97.2 | 1.7% | 265 | 374 | 3.5% | 576 | 813 | 3.5% | 3.2 | 3.9 | 7.0 | 8.4 |
| Russia | n.a. | 145.6 | n.a. | n.a. | 357 | n.a. | n.a. | 1111 | n.a. | n.a. | 2.5 | n.a. | 7.6 |

a.a.g.r. = average annual growth rate
Source: IEA.

| | Energy Production (Mtoe) | | | Share of Gas in Energy Production | | Total Primary Energy Supply (Mtoe) | | | Share of Gas in Primary Energy Supply | | Electricity Output (TWh) | | | Share of Gas in Electricity Output | |
|---|---|---|---|---|---|---|---|---|---|---|---|---|---|---|---|
| | 1990 | 2000 | a.a.g.r. 1990-2000 | 1990 | 2000 | 1990 | 2000 | a.a.g.r. 1990-2000 | 1990 | 2000 | 1990 | 2000 | a.a.g.r. 1990-2000 | 1990 | 2000 |
| **SOUTH AMERICA** | | | | | | | | | | | | | | | |
| Argentina | 47.4 | 81.2 | 5.5% | 36% | 42% | 45.0 | 61.5 | 3.2% | 36% | 42% | 51.0 | 89.0 | 5.7% | 44% | 80% |
| Bolivia | 4.9 | 5.9 | 1.8% | 56% | 53% | 2.8 | 4.9 | 5.9% | 56% | 53% | 2.1 | 4.0 | 6.4% | 29% | 37% |
| Brazil | 97.1 | 142.1 | 3.9% | 2% | 4% | 132.5 | 183.2 | 3.3% | 2% | 4% | 222.8 | 349.2 | 4.6% | 0% | 1% |
| Chile | 7.6 | 8.3 | 0.8% | 19% | 21% | 13.6 | 24.4 | 6.0% | 19% | 21% | 18.4 | 41.3 | 8.4% | 2% | 37% |
| Colombia | 48.4 | 74.6 | 4.4% | 7% | 7% | 25.0 | 28.8 | 1.4% | 7% | 7% | 36.2 | 44.0 | 2.0% | 18% | 29% |
| Ecuador | 16.1 | 22.5 | 3.4% | 1% | 1% | 5.8 | 8.2 | 3.5% | 1% | 1% | 6.4 | 10.6 | 5.2% | – | – |
| Paraguay | 4.6 | 6.9 | 4.2% | – | – | 3.1 | 3.9 | 2.4% | – | – | 27.2 | 53.5 | 7.0% | – | – |
| Peru | 10.7 | 9.5 | -1.2% | 5% | 6% | 10.1 | 12.7 | 2.4% | 5% | 6% | 13.8 | 19.9 | 3.7% | 2% | 6% |
| Trinidad & Tobago | 12.6 | 17.9 | 3.6% | 37% | 62% | 5.8 | 8.7 | 4.1% | 37% | 62% | 3.6 | 5.5 | 4.4% | 61% | 63% |
| Uruguay | 1.1 | 1.0 | -1.1% | 0% | 0% | 2.3 | 3.1 | 3.2% | 0% | 0% | 7.4 | 7.6 | 0.2% | – | – |
| Venezuela | 149.8 | 225.5 | 4.2% | 14% | 12% | 44.9 | 59.3 | 2.8% | 14% | 12% | 59.3 | 85.2 | 3.7% | 35% | 24% |
| **Total South America** | **400.5** | **595.3** | **4.0%** | **13%** | **15%** | **290.9** | **398.6** | **3.2%** | **13%** | **15%** | **448.2** | **709.7** | **4.7%** | **15%** | **23%** |
| **For comparison:** | | | | | | | | | | | | | | | |
| **IEA COUNTRIES** | | | | | | | | | | | | | | | |
| Canada | 274 | 375 | 3.2% | 32% | 40% | 209 | 251 | 1.8% | 32% | 40% | 482 | 605 | 2.3% | 5% | 13% |
| Netherlands | 60 | 57 | -0.5% | 91% | 91% | 66 | 76 | 1.3% | 91% | 91% | 72 | 90 | 2.2% | 55% | 68% |
| Norway | 120 | 225 | 6.5% | 20% | 20% | 21 | 26 | 1.8% | 20% | 20% | 122 | 142 | 1.6% | 0% | 1% |
| UK | 208 | 273 | 2.7% | 20% | 36% | 212 | 233 | 0.9% | 20% | 36% | 317 | 372 | 1.6% | 2% | 63% |
| US | 1650 | 1676 | 0.2% | 25% | 27% | 1927 | 2300 | 1.8% | 25% | 27% | 3182 | 4004 | 2.3% | 20% | 27% |
| IEA North America | 1924 | 2051 | 0.6% | 26% | 29% | 2136 | 2551 | 1.8% | 26% | 29% | 3663 | 4609 | 2.3% | 18% | 26% |
| IEA Europe | 919 | 1070 | 1.5% | 18% | 22% | 1498 | 1655 | 1.0% | 18% | 22% | 2466 | 3013 | 2.0% | 11% | 31% |
| IEA Asia-Pacific | 268 | 387 | 3.8% | 9% | 9% | 633 | 847 | 3.0% | 9% | 9% | 1145 | 1622 | 3.5% | 31% | 36% |
| **IEA Total** | **3110** | **3508** | **1.2%** | **22%** | **25%** | **4267** | **5053** | **1.7%** | **22%** | **25%** | **7275** | **9244** | **2.4%** | **18%** | **29%** |
| **OTHER COUNTRIES** | | | | | | | | | | | | | | | |
| China | 903 | 1108 | 2.1% | 2% | 3% | 870 | 1142 | 2.8% | 2% | 3% | 621 | 1356 | 8.1% | 0% | 1% |
| India | 333 | 422 | 2.4% | 3% | 5% | 359 | 502 | 3.4% | 3% | 5% | 289 | 542 | 6.5% | 3% | 5% |
| Indonesia | 162 | 229 | 3.6% | 24% | 26% | 93 | 146 | 4.6% | 24% | 26% | 37 | 93 | 9.6% | 2% | 22% |
| Mexico | 194 | 230 | 1.7% | 12% | 14% | 124 | 154 | 2.2% | 12% | 14% | 123 | 204 | 5.2% | 10% | 26% |
| Russia | 1270 | 967 | -2.7% | 41% | 49% | 868 | 614 | -3.4% | 41% | 49% | 1082 | 876 | -2.1% | 59% | 60% |

a.a.g.r. = average annual growth rate
Source: IEA.

| | Total Final Energy Consumption (Mtoe) | | | Share of Gas in Final Energy Consumption | | Total Energy Consumption in Industry (Mtoe) | | | Share of Gas in Industrial Energy Consumption | | Electricity Consumption (TWh) | | |
|---|---|---|---|---|---|---|---|---|---|---|---|---|---|
| | 1990 | 2000 | a.a.g.r. 1990-2000 | 1990 | 2000 | 1990 | 2000 | a.a.g.r. 1990-2000 | 1990 | 2000 | 1990 | 2000 | a.a.g.r. 1990-2000 |
| **SOUTH AMERICA** | | | | | | | | | | | | | |
| Argentina | 45.0 | 61.5 | 3.2% | 21% | 26% | 9.9 | 14.8 | 4.1% | 10% | 11% | 42.5 | 78.8 | 6.4% |
| Bolivia | 2.8 | 4.9 | 5.9% | 6% | 7% | 0.4 | 0.8 | 6.7% | 6% | 6% | 1.8 | 3.3 | 6.1% |
| Brazil | 132.5 | 183.2 | 3.3% | 2% | 3% | 47.2 | 66.9 | 3.5% | 1% | 2% | 217.7 | 329.8 | 4.2% |
| Chile | 13.6 | 24.4 | 6.0% | 2% | 4% | 3.7 | 6.8 | 6.3% | 0% | 3% | 16.4 | 38.3 | 8.8% |
| Colombia | 25.0 | 28.8 | 1.4% | 4% | 6% | 5.3 | 7.0 | 2.8% | 3% | 3% | 28.7 | 33.5 | 1.6% |
| Ecuador | 5.8 | 8.2 | 3.5% | 0% | 0% | 1.3 | 1.3 | 0.4% | – | – | 5.0 | 8.4 | 5.4% |
| Paraguay | 3.1 | 3.9 | 2.4% | – | – | 1.0 | 1.4 | 3.7% | – | – | 2.1 | 4.9 | 8.6% |
| Peru | 10.1 | 12.7 | 2.4% | 1% | 0% | 1.9 | 3.0 | 4.7% | – | – | 11.9 | 17.6 | 4.0% |
| Trinidad & Tobago | 5.8 | 8.7 | 4.1% | 43% | 44% | 2.8 | 4.2 | 4.2% | 43% | 44% | 3.3 | 5.1 | 4.5% |
| Uruguay | 2.3 | 3.1 | 3.2% | 0% | 1% | 0.5 | 0.5 | -1.0% | 0% | 1% | 3.9 | 6.6 | 5.4% |
| Venezuela | 44.9 | 59.3 | 2.8% | 19% | 19% | 12.9 | 15.6 | 2.0% | 17% | 16% | 48.6 | 64.5 | 2.9% |
| **Total South America** | **290.9** | **398.6** | **3.2%** | **8%** | **10%** | **86.8** | **122.4** | **3.5%** | **6%** | **7%** | **381.9** | **590.8** | **4.5%** |
| **For comparison:** | | | | | | | | | | | | | |
| **IEA COUNTRIES** | | | | | | | | | | | | | |
| Canada | 209 | 251 | 1.8% | 21% | 22% | 58 | 71 | 2.1% | 10% | 10% | 448 | 522 | 1.5% |
| Netherlands | 66 | 76 | 1.3% | 35% | 30% | 20 | 21 | 0.3% | 13% | 11% | 78 | 104 | 3.0% |
| Norway | 21 | 26 | 1.8% | 0% | 2% | 7 | 8 | 1.5% | – | 2% | 99 | 113 | 1.3% |
| UK | 212 | 233 | 0.9% | 20% | 24% | 39 | 41 | 0.6% | 6% | 7% | 307 | 358 | 1.6% |
| US | 1927 | 2300 | 1.8% | 16% | 14% | 341 | 359 | 0.5% | 6% | 5% | 2923 | 3813 | 2.7% |
| IEA North America | 2136 | 2551 | 1.8% | 16% | 15% | 399 | 430 | 0.8% | 7% | 5% | 3370 | 4335 | 2.5% |
| IEA Europe | 1498 | 1655 | 1.0% | 13% | 16% | 350 | 373 | 0.6% | 6% | 7% | 2341 | 2844 | 2.0% |
| IEA Asia-Pacific | 633 | 847 | 3.0% | 4% | 6% | 173 | 226 | 2.7% | 2% | 3% | 1099 | 1564 | 3.6% |
| **IEA Total** | **4267** | **5053** | **1.7%** | **13%** | **14%** | **922** | **1029** | **1.1%** | **6%** | **5%** | **6810** | **8742** | **2.5%** |
| **OTHER COUNTRIES** | | | | | | | | | | | | | |
| China | 870 | 1142 | 2.8% | 1% | 2% | 272 | 311 | 1.3% | 1% | 1% | 580 | 1253 | 8.0% |
| India | 359 | 502 | 3.4% | 1% | 2% | 51 | 93 | 6.2% | 1% | 2% | 238 | 399 | 5.3% |
| Indonesia | 93 | 146 | 4.6% | 6% | 6% | 13 | 23 | 5.9% | 6% | 5% | 31 | 82 | 10.1% |
| Mexico | 124 | 154 | 2.2% | 11% | 7% | 35 | 32 | -0.8% | 11% | 6% | 106 | 177 | 5.2% |
| Russia | 868 | 614 | -3.4% | 16% | 19% | 136 | 149 | 0.9% | 6% | 7% | 990 | 762 | -2.6% |

a.a.g.r. = average annual growth rate
Source: IEA.

| | Per Capita Primary Energy Supply (toe) | | Per Capita Final Energy Consumption (toe) | | Per Capita Electricity Consumption (MWh) | | TPES/GDP (toe per thousand 1995 US$ using ex. rates) | | TPES/GDP (toe per thousand 1995 US$ using PPPs) | |
|---|---|---|---|---|---|---|---|---|---|---|
| | 1990 | 2000 | 1990 | 2000 | 1990 | 2000 | 1990 | 2000 | 1990 | 2000 |
| **SOUTH AMERICA** | | | | | | | | | | |
| Argentina | 1.38 | 1.66 | 1.38 | 1.66 | 1.3 | 2.1 | 0.24 | 0.21 | 0.17 | 0.14 |
| Bolivia | 0.42 | 0.59 | 0.42 | 0.59 | 0.3 | 0.4 | 0.51 | 0.62 | 0.21 | 0.26 |
| Brazil | 0.90 | 1.07 | 0.90 | 1.07 | 1.5 | 1.9 | 0.22 | 0.23 | 0.14 | 0.15 |
| Chile | 1.04 | 1.60 | 1.04 | 1.60 | 1.3 | 2.5 | 0.32 | 0.30 | 0.19 | 0.18 |
| Colombia | 0.72 | 0.68 | 0.72 | 0.68 | 0.8 | 0.8 | 0.34 | 0.30 | 0.13 | 0.12 |
| Ecuador | 0.57 | 0.65 | 0.57 | 0.65 | 0.5 | 0.7 | 0.38 | 0.45 | 0.19 | 0.22 |
| Paraguay | 0.73 | 0.71 | 0.73 | 0.71 | 0.5 | 0.9 | 0.40 | 0.42 | 0.17 | 0.17 |
| Peru | 0.47 | 0.49 | 0.47 | 0.49 | 0.6 | 0.7 | 0.24 | 0.21 | 0.13 | 0.11 |
| Trinidad & Tobago | 4.77 | 6.66 | 4.77 | 6.66 | 2.7 | 3.9 | 1.17 | 1.30 | 0.71 | 0.79 |
| Uruguay | 0.72 | 0.92 | 0.72 | 0.92 | 1.2 | 2.0 | 0.15 | 0.15 | 0.11 | 0.11 |
| Venezuela | 2.30 | 2.45 | 2.30 | 2.45 | 2.5 | 2.7 | 0.69 | 0.74 | 0.41 | 0.44 |
| **Total South America** | **0.99** | **1.15** | **0.99** | **1.15** | **1.3** | **1.7** | **0.27** | **0.27** | **0.16** | **0.17** |
| **For comparison:** | | | | | | | | | | |
| **IEA COUNTRIES** | | | | | | | | | | |
| Canada | 7.55 | 8.16 | 7.55 | 8.16 | 16.2 | 17.0 | 0.39 | 0.36 | 0.34 | 0.31 |
| Netherlands | 4.45 | 4.76 | 4.45 | 4.76 | 5.2 | 6.6 | 0.18 | 0.15 | 0.22 | 0.19 |
| Norway | 5.06 | 5.70 | 5.06 | 5.70 | 23.4 | 25.2 | 0.18 | 0.15 | 0.25 | 0.22 |
| UK | 3.69 | 3.89 | 3.69 | 3.89 | 5.3 | 6.0 | 0.20 | 0.18 | 0.21 | 0.18 |
| US | 7.71 | 8.35 | 7.71 | 8.35 | 11.7 | 13.8 | 0.30 | 0.26 | 0.30 | 0.26 |
| IEA North America | 7.69 | 8.33 | 7.69 | 8.33 | 12.1 | 14.2 | 0.30 | 0.26 | 0.30 | 0.26 |
| IEA Europe | 3.30 | 3.47 | 3.30 | 3.47 | 5.2 | 6.0 | 0.17 | 0.16 | 0.20 | 0.18 |
| IEA Asia-Pacific | 3.39 | 4.30 | 3.39 | 4.30 | 5.9 | 7.9 | 0.11 | 0.12 | 0.18 | 0.20 |
| **IEA Total** | **4.65** | **5.15** | **4.65** | **5.15** | **7.4** | **8.9** | **0.20** | **0.19** | **0.23** | **0.22** |
| **OTHER COUNTRIES** | | | | | | | | | | |
| China | 0.77 | 0.90 | 0.77 | 0.90 | 0.5 | 1.0 | 2.20 | 1.10 | 0.48 | 0.24 |
| India | 0.42 | 0.49 | 0.42 | 0.49 | 0.3 | 0.4 | 1.31 | 1.08 | 0.27 | 0.22 |
| Indonesia | 0.52 | 0.69 | 0.52 | 0.69 | 0.2 | 0.4 | 0.67 | 0.70 | 0.24 | 0.25 |
| Mexico | 1.52 | 1.58 | 1.52 | 1.58 | 1.3 | 1.8 | 0.47 | 0.41 | 0.22 | 0.19 |
| Russia | n.a. | 4.22 | n.a. | 4.22 | n.a. | 5.2 | n.a. | 1.72 | n.a. | 0.55 |

Source: IEA.

# ABBREVIATIONS

| | |
|---|---|
| a.a.g.r. | average annual growth rate |
| ALADI | *Latin American Integration Association* |
| ANEEL | *Agência Nacional de Energia Elétrica* (National Electricity Agency), Brazil |
| ANP | *Agência Nacional do Petróleo* (National Petroleum Agency), Brazil |
| $B | boliviano (Bolivian currency) |
| bcf | billion cubic feet |
| bcf/d | billion cubic feet per day |
| bcm | billion cubic meter |
| bcm/a | billion cubic meters per annum |
| bcm/d | billion cubic meters per day |
| BG | *British Gas* |
| BNDES | *Banco Nacional de Desenvolvimento Econômico e Social* (Brazilian Central Development Bank), Brazil |
| BOO | build-own-operate |
| bpd | barrels per day |
| Btu | British thermal unit |
| CAF | *Corporación Andina de Fomento* (Andean Development Corporation) |
| CAN | *Comunidad Andina* (Andean Community) |
| CCGT | combined-cycle gas turbines |
| CEG | *Companhia Estadual de Gás,* Brazil |
| CEPAL | *Comisión Económica para América Latina y el Caribe de las Naciones Unidas* (United Nations Economic Commission for Latin America and the Caribbean) |
| CIESA | *Compañía de Inversiones de Energía S.A.,* Brazil |
| cm | cubic metre |
| CNEA | *Comisión Nacional de Energía Atómica* (National Commission for Atomic Energy), Argentina |
| CNG | compressed natural gas |
| CNPE | *Conselho Nacional de Politica Energetica* (National Council for Energy Policy), Brazil |
| CO$_2$ | carbon dioxide |
| CSPE | *Comissão de Serviços Públicos de Energia* (São Paulo Regulatory Agency for Electricity and Gas), Brazil |
| DOMEGAS | *Venezolana Domestica de Gas,* Venezuela |
| ECLAC | see UN-ECLAC |
| E&P | exploration and production |
| EFH | *Ente Federal de Hidrocarburos,* Argentina |
| EIU | *The Economist Intelligence Unit* |
| Enagas | *Ente Nacional del Gas* (National Gas Agency), Venezuela |

| | |
|---|---|
| Enargas | *Ente Nacional Regulador del Gas* (National Regulatory Agency for Gas), Argentina |
| ENRE | *Ente Nacional Regulador de la Electricidad* (National Regulatory Agency for Electricity), Argentina |
| FDI | foreign direct investment |
| FTAA | Free-Trade Agreement of the Americas |
| Gasbol | Bolivia-Brazil pipeline |
| GdE | *Gas del Estado*, Argentina |
| GDP | gross domestic product |
| GTB | *Gas Transboliviano Ltda,* Bolivia |
| GTL | gas-to-liquids |
| GSA | gas supply agreement |
| GW | gigawatt |
| HSE | health, safety and the environment |
| IBGE | *Instituto Brasileiro de Geografia e Estatística*, Brazil |
| IDB | *Inter-American Development Bank* |
| IEA | *International Energy Agency* |
| IMF | *International Monetary Fund* |
| IPP | independent power producer |
| IRR | internal rate of return |
| JV | joint venture |
| kb/a | thousand barrels per day |
| kcal | kilocalorie |
| kg | kilogramme |
| km | kilometre |
| km² | square kilometre |
| ktoe | thousand tonnes of oil equivalent |
| kW | kilowatt |
| kWh | kilowatt-hour |
| LDC | local distribution companies |
| LNG | liquefied natural gas |
| LPG | liquefied petroleum gas |
| MAE | *Mercado Atacadista de Energia* (Wholesale Electricity Market), Brazil |
| Mb/d | million barrels per day |
| MBtu | million of British thermal units |
| Mcf | million cubic feet |
| mcm | million cubic metres |
| mcm/d | million cubic metres per day |
| MCT | *Ministerio da Ciencia e Tecnologia* (Ministry of Science and Technology), Brazil |
| MEM | *Ministerio de Energía y Minas* (Ministry of Energy and Mines), various countries |
| Mercosur | *Mercado Común del Sur* (Southern Common Market – in Spanish) |
| Mercosul | *Mercado Común do Sul* (Southern Common Market – in Portuguese) |
| MMA | *Ministerio de Medio Ambiente* (Ministry of the Environment), various countries |
| MME | *Ministerio de Minas e Energia* (Ministry of Mines and Energy), Brazil |

| | |
|---|---|
| Mtoe | million tonnes of oil equivalent |
| MW | megawatt |
| MWh | megawatt-hour |
| NAFTA | North American Free Trade Agreement |
| NGL | natural gas liquids |
| NGV | natural-gas vehicle |
| NOx | nitrogen oxide |
| OAS | *Organisation of the American States* |
| OECD | *Organisation for Economic Co-operation and Development* |
| OLADE | *Organisación Latinoaméricana de Energía* (Latin American Energy Organisation) |
| ONS | *Operador Nacional do Sistema Elétrico* (National Electric System Operator), Brazil |
| p.a. | per annum |
| PDVSA | *Petróleos de Venezuela S.A.*, Venezuela |
| PND | *Programa Nacional de Desestatização,* Brazil |
| PPA | power purchase agreements |
| PPP | purchasing power parity |
| PPT | *Programa Prioritário de Termelétricas* (Thermoelectric Priority Programme), Brazil |
| R&D | research and development |
| ROR | rate of return |
| R/P | reserve-to-production |
| R$ | real, reais (Brazilian currency) |
| SH | *Superintendencia de Hidrocarburos* (Superintendency of Hydrocarbons), Bolivia |
| SIRESE | *Sistema de Regulacion Sectorial* (Sectorial Regulatory System), Bolivia |
| SIC | *Sistema Interconectado Central* (Central Interconnected System), Chile |
| $SO_2$ | sulphur dioxide |
| SOE | state-owned enterprise |
| T&D | transmission & distribution |
| tcf | trillion cubic feet |
| tcm | trillion cubic metres |
| TBG | *Transportadora Brasileira Gasoduto Bolívia-Brasil S.A.*, Brazil |
| TGM | *Transportadora de Gas de Mercosur* |
| TGN | *Transportadora de Gas del Norte,* Argentina |
| TGS | *Transportadora de Gas del Sur,* Argentina |
| TJ | terajoule |
| toe | tonnes of oil equivalent |
| TPA | third party access |
| TPES | total primary energy supply |
| TSB | *Transportadora Sul Brasileira de Gas,* Brazil |
| TWh | terrawatt-hour |
| UK | United Kingdom |
| UNCTAD | *United Nations Conference on Trade and Development* |
| UN-ECLAC | *United Nations Economic Commission for Latin America and the Caribbean* |
| US | United States of America |

| | |
|---|---|
| US DOE | *US Department of Energy* |
| US$ | American dollar |
| USGS | US Geological Survey |
| VAT | value added tax |
| VDGAS | *Venezolana Distribudora de Gas Natural*, Venezuela |
| VNG | vehicular natural gas |
| WB | World Bank |
| WTO | World Trade Organisation |
| YPF | *Yacimientos Petrolíferos Fiscales S.A.,* Argentina |
| YPFB | *Yacimientos Petrolíferos Fiscales Bolivianos S.A.*, Bolivia |

# GLOSSARY

### Associated gas
Gas reserves found in reservoirs that hold a large percentage of liquid hydrocarbons.

### Biomass
Biomass energy, or bioenergy, is defined as any plant matter used directly as fuel or converted into other forms before combustion. It includes wood, vegetal waste (including wood waste and crops used for energy production), animal wastes, sulphite lyes, also known as "black liquor", and other solid biomass.

### City gate
Point at which the local distribution company takes delivery of gas; physical interface between transmission and local distribution systems.

### Capacity charge
Price asked for reservation of particular capacity of the gas infrastructure (e.g. pipeline/s, storage), independent of whether used or not.

### Captive customers
A consumer who does not have any choice but to use gas, or a consumer who, once connected to gas, has no or little ability to switch to another fuel in view of the high costs that such switch would imply.

### Commodity charge
Price asked for the gas volume as such (for gas as a commodity; distinct from other charged costs such as customer charge, capacity charge or charge for other services).

### Cost-plus pricing
A pricing approach according to which the price of a commodity or a service is based on its costs of production regardless of market conditions. See *Chapter 4* for a more in-depth discussion.

### Downstream
Segment of the natural gas chain that includes distribution and sales to large gas users. Sometimes also used to include gas transmission, as opposed to *upstream*.

### Economies of scale
The reduction in unit cost when producing larger quantities of one product.

### Economies of scope
Cost savings due to synergies between different activities (e.g. combined meter reading and billing in mixed energy utilities supplying gas, electricity and water).

### Feedstock
Natural gas which is used as essential component of a chemical product, e.g. fertilizer.

### Free gas
See *Non-associated gas*.

### Gross/net calorific basis
The difference between the "net" and the "gross" calorific value is the latent heat of vaporisation of the water vapour produced during combustion if the fuel. For natural gas, the net calorific value is 10 per cent less than the gross. In IEA statistics and in this report, gas data presented in bcm are usually on a gross calorific basis, while gas data presented in Mtoe (mainly for comparisons with other fuels) are on a net calorific basis.

### Gross gas production
Gross production corresponds to the total production of natural gas, before flaring, re-injection, losses and processing.

### Indigenous gas production
Defined in IEA statistics as all dry marketable production within national boundaries, including offshore production. Production is measured after purification and extraction of NGLs and sulphur. Quantities reinjected, vented and flared, are not included. Corresponds to the concept of *Marketed gas production*.

### Interruptible service
Gas sales and transportation service that is offered at a rebate to compensate for the right to interrupt the delivery of gas or the service according to previously agreed criteria (e.g. cold weather, or at discretion, but only for a total limited time).

### Load factor
Ratio that the average demand or throughput over a year bears to the maximum demand for a given time period (e.g. one day, one hour, etc.) or to the capacity expressed as a percentage.

### Market replacement value
The price of gas at which the end-user incurs the same costs when using gas instead of alternative fuels such as coal and oil products, taking into account differences in heat value, efficiency and costs for the appliances or equipment necessary to use the energy, plus eventually different valuation of resulting pollution, if provided by a market mechanism. See *Chapter 4* for a more in-depth discussion.

### Marketed gas production
Marketed production is the concept adopted at the international level to assess the contribution of the "natural gas" primary energy in the energy balances. Production is measured after purification and extraction of NGL and sulphur, and excludes reinjected gas and quantities vented or flared. It includes gas consumed by gas processing plants and gas transported by pipeline. Corresponds in IEA gas statistics to *Indigenous gas production*.

### Midstream
Segment of the natural gas chain that includes transmission and storage. Sometimes included into *downstream*, as opposed to *upstream*.

### Natural monopoly
An economic activity in which the production of a particular good or service by a single producer results in lower unit cost than by multiple producers. It usually involves an

industry with a decreasing marginal cost of production and high entry barriers, so that the incumbent producer can monopolise the market.

### Net-back pricing

Delivered price of cheapest alternative fuel to gas to the customer (including any taxes) adjusted for any efficiency differences in the energy conversion process,

- *minus* cost of transporting gas from the beach/border to the customer;
- *minus* cost of storing gas to meeting seasonal or daily demand fluctuations;
- *minus* gas taxes.

   See *Chapter 4* for a more in-depth discussion.

### Net calorific basis

See *Gross/net calorific basis.*

### Non-associated gas

Gas reserves found in reservoirs that contain mostly gaseous hydrocarbons.

### Off-take

Actual amount of gas taken from the transmission or distribution system.

### Opportunity cost

The net economic loss (forgone earnings minus forgone costs) by choosing one opportunity over another.

### Power Purchase Agreement

A contract entered into by an independent power producer and an electric utility, which specifies the terms and conditions under which electric power will be generated and purchased. The PPA require the independent power producer to supply power at a specified price for the life of the agreement. PPA contracts normally include: specification of the size and operating parameters of the generating facility; milestones, in-service dates and contract terms; price mechanisms; service and performance obligations; dispatchability options; and conditions of termination or default.

### Peak storage

Gas storage designed/used to supplement normal gas supply during short periods of extremely high demand.

### Possible reserves

Reserves estimated in reservoirs that have been tentatively identified in undrilled areas adjacent to proven or probable geological volumes. The assessment of these reserves depends on assumptions concerning the geometry and impregnation of these reservoirs.

### Probable reserves

Reserves that geological and engineering information indicates with good probability can be recovered in the future from known reservoirs under economic and technical conditions similar to those of proved reserves. These reserves are roughly measured and their reservoirs are not equipped to produce.

### Proven reserves

Reserves that geological and engineering information indicates with reasonable certainty can be recovered in the future from known reservoirs under existing economic and operating conditions. These reserves are located in fully explored reservoirs, already equipped or currently being equipped to produce.

### Rent

The difference between the market replacement value and a lower price of gas paid by the consumer (consumer rent), or the difference between the price paid to the producer and the cost of production incurred by the producer (producer rent).

### Take-or-Pay (TOP)

A contractual commitment on the part of a buyer to take a minimum volume of gas or pay for it if not taken, usually over a 12-month period.

### Third party access (TPA)

The obligation to the owner of an essential facility to offer access to third parties to idle capacity of such facility under non-discriminatory conditions.

### Total final consumption (TFC)

Total energy final consumption is defined in IEA statistics and balances as the sum of energy consumption by the different end-use sectors: industry, transport, agriculture, commercial and public services, and residential.

### Total primary energy supply (TPES)

Total primary energy supply is defined in IEA statistics and balances as made up of production + imports - exports - international marine bunkers ± stock changes. It is also equal to total final energy consumption plus the energy consumed for energy transformation (electricity production, petroleum refineries, coal transformation, etc) including own use and distribution losses.

### Upstream

Segment of the natural gas chain that includes exploration, development, production, gathering and purification. The term is used in opposition to *downstream*, or to *midstream* and *downstream*.

### Volume shrinkage

Losses essentially resulting from the processing of natural gas before it enters the networks, purification and/or extraction of its liquefiable fractions (ethane, LPG, natural gasoline and condensates): this volume shrinkage may include energy self-consumption corresponding to the processing operations.

## GEOGRAPHICAL COVERAGE

**South America (as defined in this report):** Argentina, Bolivia, Brazil, Chile, Colombia, Ecuador, Paraguay, Peru, Uruguay, Trinidad and Tobago, Venezuela.

**Andean Community member countries:** Bolivia, Colombia, Ecuador, Peru and Venezuela.

**Mercosur member countries:** Argentina, Brazil, Paraguay and Uruguay, with Bolivia and Chile as associate members.

**Southern Cone:** a geographical area encompassing Argentina, Bolivia, Chile, Paraguay and Uruguay, as well as the southern half of Brazil.

**OECD member countries:** Australia, Austria, Belgium, Canada, the Czech Republic, Denmark, Finland, France, Germany, Greece, Hungary, Iceland, Ireland, Italy, Japan, Luxembourg, Mexico, the Netherlands, New Zealand, Norway, Poland, Portugal, Republic of Korea, Slovakia, Spain, Sweden, Switzerland, Turkey, the United Kingdom, and the United States.

**IEA member countries:** All OECD member countries except Iceland, Mexico, Poland and Slovakia.

**IEA Europe:** Austria, Belgium, the Czech Republic, Denmark, Finland, France, Germany, Greece, Hungary, Ireland, Italy, Luxembourg, the Netherlands, Norway, Portugal, Spain, Sweden, Switzerland, Turkey, and the United Kingdom.

**IEA North America:** Canada and the United States.

**IEA Asia-Pacific:** Australia, Japan, New Zealand and Republic of Korea.

# BIBLIOGRAPHY

ALMEIDA E. and J. MACHADO (2001),
"Mercosul: A Nova Integração Energética" in *El Desafío de Integrarse para Crecer: Balance y Perspectivas del Mercosur en su Primera Década,* D. Chudnovsky and J.M. Fanelli (eds), Buenos Aires.

ANP (2001a),
*Anuario Estatístico da la Indústria Brasileira do Petróleo 1991-2000,* ANP, Rio de Janeiro.

*ANP (2001b),*
Indústria Brasileira de Gás Natural: Regulação Atual e Desafios Futuros, *Séries ANP, Número 2, Rio de Janeiro.*

ANP (2002a),
"Panorama da Indústria de Gás Natural no Brasil: Aspectos Regulatórios e Desafios", Nota Técnica 033/2002-SCG.

ANP (2002b),
"Participações Cruzadas na Indústria Brasileira de Gás Natural", SCG/ANP, February.

BEATO P. and C. FUENTE (2000),
"Liberalization of the Gas Sector in Latin America: The Experience of Three Countries", Inter-American Development Bank, Sustainable Development Department, Best Practices Series, June.

BP (2002),
*BP Statistical Review of World Energy 2000,* www.bp.com/centres/energy2002/index.asp

CAMPODONICO H. (1998),
"La Industria del Gas Natural y las Modalidades de Regulación en América Latina", CEPAL, Serie Medio Ambiente y Desarrollo, No. LC/L 1121, April.

CCPE (2001),
*Plano Decenal de Expansão 2001-2010,* Brasilia.

CEDIGAZ (2002),
*Natural Gas in the World – 2002 Survey,* Paris.

CEPAL (2001),
*Anuario Estadístico de América Latina y el Caribe,* LC/G. 2151-P/B, Santiago de Chile.

CEPAL (2002),
*Economic Survey of Latin America and the Caribbean 2001-2002,* LC/G. 2184-P, Santiago de Chile.

CEPAL (2001),
*Inversión Extranjera en América Latina y el Caribe,* LC/G. 2178-P, Santiago de Chile.

CINTRA T. (2002),
"Electricity and Gas in the Southern Cone of Latin America: Can Market Integration Succeed?", Degree Thesis, University of Sheffield.

DOS SANTOS E. et al. (2000),
"Development of the Brazilian gas sector and the role of power generation: the difficult choice between 'flaring gas' or 'bubbling water'", paper presented at the Conference on Oil and Gas Investments in Brazil, New York, 28-29 September.

EIU – The Economist Intelligence Unit (2002a),
*Argentina – Country Profile 2002,* www.eiu.com

EIU – The Economist Intelligence Unit (2002b),
*Brazil – Country Profile 2002,* www.eiu.com

EIU – The Economist Intelligence Unit (2002c),
*Brazil – Country Profile 2002,* www.eiu.com

FERNANDEZ E. and J. PERIN SILVEIRA (1999),
"A Reforma do Setor Petrolífero na América Latina: Argentina, México e Venezuela", ANP, March.

FERNANDEZ M. and E. BIRHUET (2002),
"Resultados de la Reestructuración Energética en Bolivia", CEPAL, Serie Recursos Naturales e Infraestructura, No. LC/L 1728-P, April.

FIGUEROA de la VEGA F. (1999),
"Interconexiones y Perspectivas para el Comercio de Gas Natural en América Latina y el Caribe 2000-2020", OLADE-CEPAL-GTZ *Working Paper*, June.

FIGUEROA de la VEGA F. (2000a),
"Energy Sector Modernization in Latin America and the Caribbean", CEPAL, November.

FIGUEROA de la VEGA F. (2000b),
"Precios des Gas Natural en América Latina: La Evidencia Empírica", OLADE-CEPAL-GTZ *Working Paper*, March.

GADANO N. (1998),
"Determinantes de la Inversión en el Sector Petróleo y Gas de la Argentina", Serie Reformas Económicas de la CEPAL, No. LC/L 1154, October.

IEA (1994),
*Natural Gas Transportation: Organisation and Regulation*, IEA/OECD, Paris.

IEA (1996),
*Regulatory Reform in Mexico's Natural Gas Sector*, IEA/OECD, Paris.

IEA (1998a),
*Natural Gas Distribution - Focus on Western Europe*, IEA/OECD, Paris.

IEA (1998b),
*Natural Gas Pricing in Competitive Markets*, IEA/OECD, Paris.

IEA (1998c),
*Biomass Energy: Data Analyses and Trends,* Proceedings, IEA/OECD, Paris.

IEA (1998c),
*World Energy Outlook 1998,* IEA/OECD, Paris.

IEA (1999),
*Regulatory Reform in Argentina's Natural Gas Sector*, IEA/OECD, Paris.

IEA (2000),
*Regulatory Reform: European Gas*, IEA/OECD, Paris.

IEA (2001),
*Distributed Generation in OECD Electricity Markets*, IEA/OECD, Paris.

IEA (2002a),
> *Energy Balances of Non-OECD Countries, 1999-2000,* IEA/OECD, Paris.

IEA (2002b),
> *Energy Balances of OECD Countries, 1999-2000,* IEA/OECD, Paris.

IEA (2002c),
> *Energy Statistics of Non-OECD Countries, 1999-2000,* IEA/OECD, Paris.

IEA (2002d),
> *Energy Statistics of OECD Countries, 1999-2000,* IEA/OECD, Paris.

IEA (2002e),
> *Flexibility in Demand and Supply of Natural Gas*, IEA/OECD, Paris.

IEA (2002f),
> *Natural Gas Information 2002*, IEA/OECD, Paris.

IEA (2002g),
> *World Energy Outlook 2002*, IEA/OECD, Paris.

JADRESIC A. (2000),
> "Investment in Natural Gas Pipelines in the Southern Cone of Latin America", World Bank, *Policy Research Working Paper*, No. 2315, April.

JOBITY R. and S. RACHA (1999),
> "The Atlantic LNG Project: the State of Play", Central Bank of Trinidad and Tobago Economic Bulletin, Volume 1, No. 2, p.64, August.

LAW P.L. and H.G. GARCIA (1999),
> "Gas Market Development in Brazil. How Gas-Fuelled Power Plants can Operate in Brazil's Hydro-Dominated System", World Bank, Energy Issues, No. 17, February.

LAW P.L. and N. de FRANCO (1998),
> "International Gas Trade – The Bolivia-Brazil Gas Pipeline", World Bank, Private Sector, No. 144, May.

MINISTERIO DE MINAS E ENERGIA (2001),
> *Plano Decenal de Expansão 2001-2010,* Brasilia.

MINISTERIO DE MINAS E ENERGIA (2002),
> *Balanço Energético Nacional 2002*, Brasilia.

OECD (2001),
> *Brazil Economic Survey*, OECD, Paris.

OLADE (1998a),
> *El Gas Natural en la Política Energética de América Latina y el Caribe*, Quito.

OLADE (1998b),
> *Energy Sector Modernization in Latin America and the Caribbean. Regulatory Framework, Divestiture, and Free Trade*, Quito.

OLADE (1999),
> *Energy Interconnections and Regional Integration in Latin America and the Caribbean*, Quito.

OLADE (2000),
> *Energy Report of Latin America and the Caribbean 1999 and Forecasting 2000-2020*, Quito.

OLADE (2001a),
> *Energy Report of Latin America and the Caribbean 2000*, Quito.

OLADE (2001b),
> *Study for Natural Gas Market Integration in South America*, Quito.

PISTONESI H. (2001),
"Desempeño de las Industrias de Electricidad y Gas Natural después de la Reformas: el Caso de Argentina", ILPES/CEPAL, Serie Gestión Pública, LC/L 1659-P, December.

POSTE D'EXPANSION ECONOMIQUE DE BUENOS AIRES (1999),
*Le Pétrole et le Gaz en Argentine*, CFCE, Paris.

POSTE D'EXPANSION ECONOMIQUE DE CARACAS (2000),
*Le Gaz Naturel au Venezuela – Cadre Institutionnel, Marché et Projets*, CFCE, Paris.

RODRIGUES PADILLA V. (2002),
"Estudio de Suministro de Gas Natural desde Venezuela y Colombia a Costa Rica y Panamá", CEPAL, Serie Recursos Naturales e Infraestructura, No. LC/L 1674-P, June.

SECRETARÍA DE ENERGÍA Y MINAS (2001),
*Prospectiva 2000*, Argentina.

US Geological Survey (2000),
*World Petroleum Assessment 2000*, Denver.

YPFB Vicepresidencia de Negociaciones Internacionales y Contratos (monthly),
*Informe Mensual*.

# WEB RESOURCES

### Argentina
Enargás: www.enargas.gov.ar
Instituto Argentino de la Energía "General Mosconi": www.iae.org.ar
Instituto Argentino del Petróleo y del Gas: www.iapg.org.ar
Secretaría de Energía: http://energia.mecon.gov.ar

### Bolivia
Cámara Boliviana de Hidrocarburos: www.cbh.org.bo
Ministerio de Hidrocarburos y Energía: www.hidrocarburos.gov.bo
Superintendencia de Hidrocarburos: www.superhid.gov.bo

### Brazil
Agência Nacional do Petróleo: www.anp.gov.br
Associação Brasileira de Agências Reguladoras: www.abar.org.br
Associação Brasileira das Empresas Distribuidoras de Gás Canalizado: www.abegas.com.br
Banco Nacional de Desenvolvimento Econômico e Social: www.bndes.gov.br
Brazilian Petroleum and Gas Institute: www.ibp.org.br
EnergiaBrazil: www.energiabrasil.gov.br
Gaspetro: www.gaspetro.com.br
Infrastructure Brazil: www.infraestruturabrasil.gov.br
Instituto Brasileiro de Geografia e Estatística: www.ibge.gov.br
Ministério de Minas e Energia: www.mme.gov.br
National Electric Agency: www.aneel.gov.br
Petrobras: www.petrobras.com.br
Portal Gás Energia: www.gasenergia.com.br
Transportadora Brasileira Gasoduto Bolívia Brasil S.A.: www.tbg.com.br

### Chile

Comisión Nacional de Energía: www.cne.cl
Empresa Nacional del Petróleo: www.enap.cl

### Colombia

Comisión de Regulación de Energía y Gas: www.creg.gov.co
Ecopetrol: www.ecopetrol.com.co
Ecogás: www.ecogas.com.co
Ministerio de Minas y Energia: www.minminas.gov.co
Naturgas: www.naturgas.com.co
Unidad de Planeación Minero Energética: www.upme.gov.co

### Peru

Camisea Project: www.camisea.com.pe
Ministerio de Energía y Minas: www.mem.gob.pe
Sociedad Nacional de Minería, Petróleo y Energía: www.snmpe.org.pe

### Trinidad and Tobago

Ministry of Energy and Energy Industries: www.energy.gov.tt
The National Gas Company of Trinidad and Tobago Ldt.: www.ngc.co.tt

### Venezuela

Cámara Petrolera de Venezuela: www.camarapetrolera.org
Enargas: www.enagas.gov.ve
PDVSA: www.pdvsa.com

### Other

Economic Commission for Latin America and the Caribbean: www.eclac.org
International Energy Agency: www.iea.org
Organisation for Economic Co-Operation and Development: www.oecd.org
OLADE: www.olade.org
US Department of Energy, Fossil Energy, Country Reports: www.fe.doe.gov
US Department of Energy, Country Briefs: http://www.eia.doe.gov/emeu/cabs/
US Department of State, Country Reports: www.state.gov

# MAPS

**Map 1**  Major natural gas basins in South America

Source: IEA.

**Map 2** South America's natural gas reserves (1 January 2002) and 2001 production

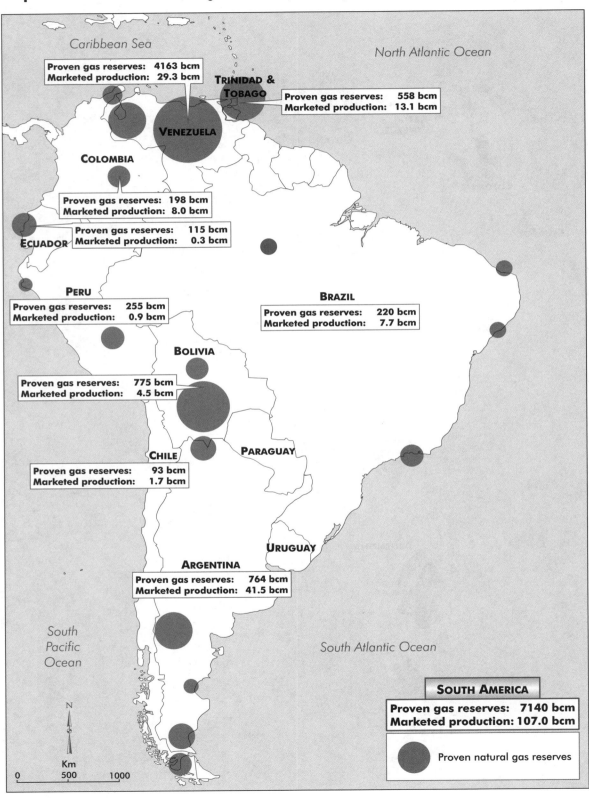

*Caribbean Sea*

*North Atlantic Ocean*

Proven gas reserves: 4163 bcm
Marketed production: 29.3 bcm

**TRINIDAD &
TOBAGO**

Proven gas reserves: 558 bcm
Marketed production: 13.1 bcm

**VENEZUELA**

**COLOMBIA**

Proven gas reserves: 198 bcm
Marketed production: 8.0 bcm

Proven gas reserves: 115 bcm
Marketed production: 0.3 bcm

**ECUADOR**

**PERU**

Proven gas reserves: 255 bcm
Marketed production: 0.9 bcm

**BRAZIL**

Proven gas reserves: 220 bcm
Marketed production: 7.7 bcm

**BOLIVIA**

Proven gas reserves: 775 bcm
Marketed production: 4.5 bcm

**CHILE**

**PARAGUAY**

Proven gas reserves: 93 bcm
Marketed production: 1.7 bcm

**URUGUAY**

**ARGENTINA**

Proven gas reserves: 764 bcm
Marketed production: 41.5 bcm

*South
Pacific
Ocean*

*South Atlantic Ocean*

N

Km
0    500    1000

**SOUTH AMERICA**

Proven gas reserves: 7140 bcm
Marketed production: 107.0 bcm

Proven natural gas reserves

*Source: IEA.*

**Map 3** Cross-border pipelines in South America, 2002

*Source: IEA.*

**Map 4** Cross-border gas flows in South America, 2001

Caribbean Sea

Caracas

**TRINIDAD & TOBAGO**

LNG exports
3.8 bcm

North Atlantic Ocean

**VENEZUELA**

Bogotá

**COLOMBIA**

Quito

**ECUADOR**

**SOUTH AMERICA**

Total gas flows in 2001:
by pipelines = 9.9 bcm
LNG exports = 3.8 bcm

**PERU**

Lima

**BRAZIL**

**BOLIVIA TO BRAZIL**
3.7 bcm

Brasília

**BOLIVIA**

La Paz

**BOLIVIA TO CHILE**
124 mcm

**ARGENTINA TO CHILE**
1.4 bcm

**PARAGUAY**
Asunción

São Paulo

**CHILE**

**ARGENTINA TO BRAZIL**
740 mcm

**URUGUAY**

Santiago

Buenos Aires

Montevideo

**ARGENTINA TO CHILE**
2.3 bcm

**ARGENTINA**

**ARGENTINA TO URUGUAY**
36 mcm

South
Pacific
Ocean

N

South Atlantic Ocean

Km
0    500    1000

**ARGENTINA TO CHILE**
1.5 bcm

▲ LNG export terminal

Source: IEA.

**Map 5**  Projected cross-border gas flows in South America, 2010

Caribbean Sea

**VENEZUELA LNG exports**

**TRINIDAD & TOBAGO LNG exports**

Caracas

**TRINIDAD & TOBAGO**

North Atlantic Ocean

**PANAMA**

Bogotá

**COLOMBIA**

**VENEZUELA**

Quito

**ECUADOR**

**Brazil LNG imports**

**PERU**

Lima

**BRAZIL**

**BOLIVIA**

La Paz

Brasília

**PERU LNG exports**

**BOLIVIA LNG exports through CHILE or PERU**

**PARAGUAY**

Asunción

**CHILE**

São Paulo

**URUGUAY**

Santiago

Buenos Aires

Montevideo

**ARGENTINA**

South Pacific Ocean

South Atlantic Ocean

N

Km
0  500  1000

▲ LNG export terminals

Source: IEA.

**Map 6**  Argentina's natural gas basins

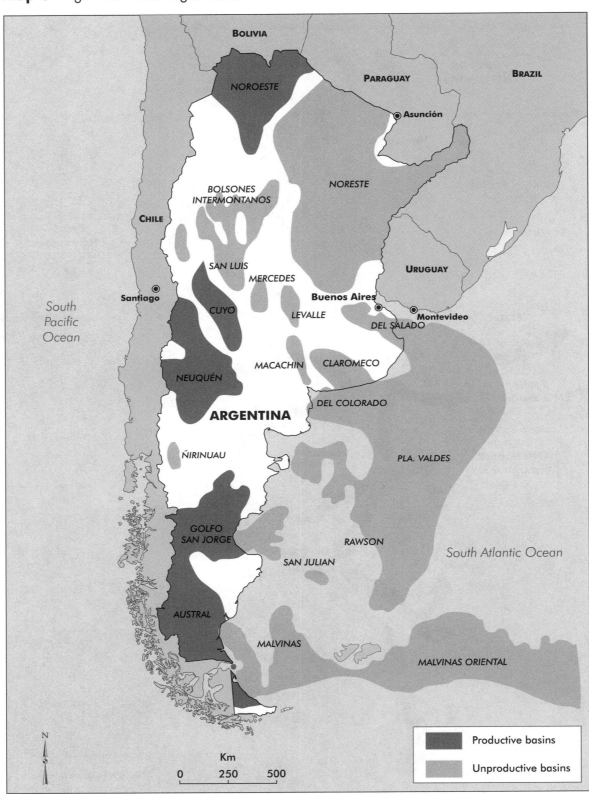

**Map 7** Argentina's gas transmission and distribution networks, 2002

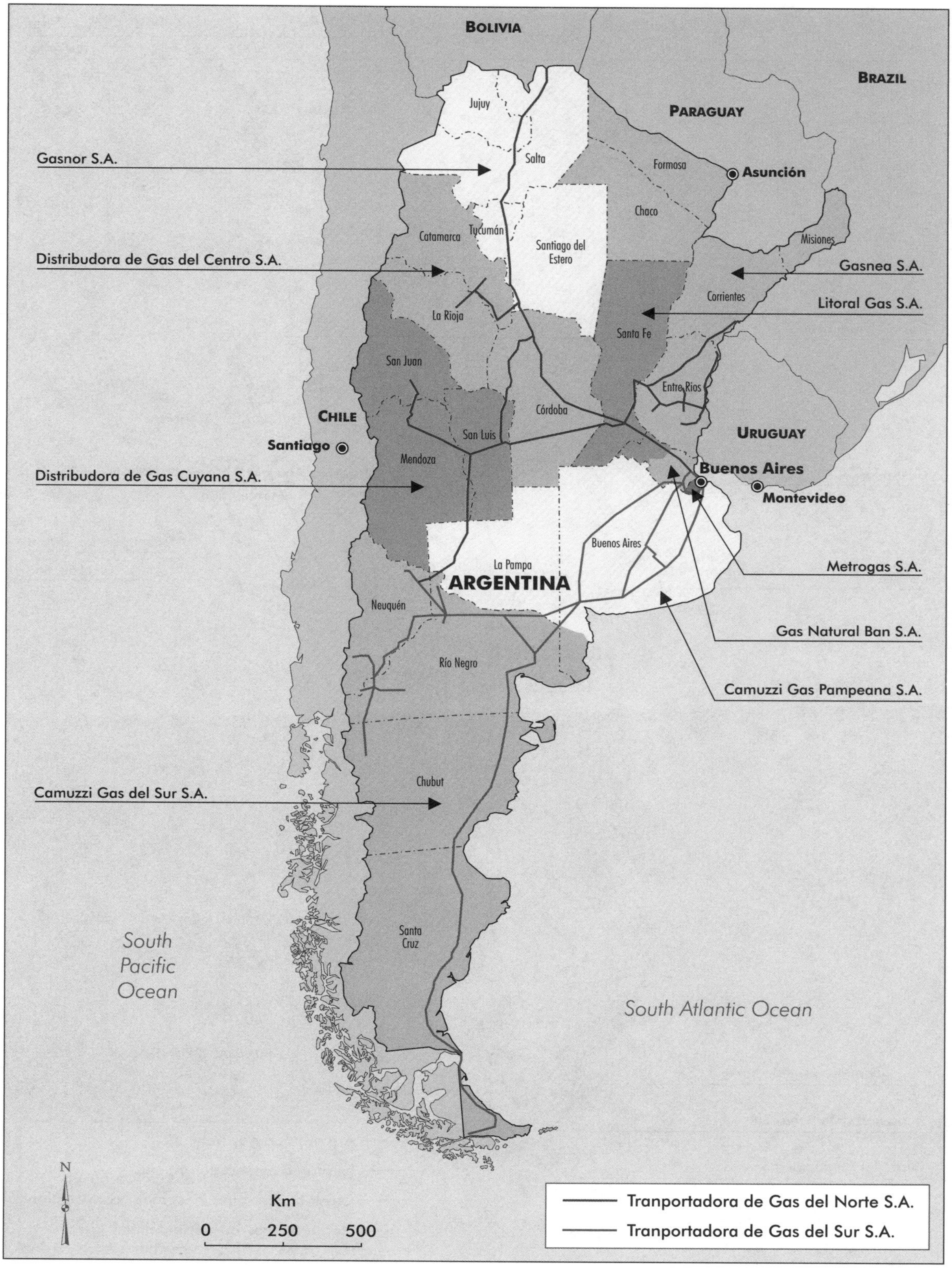

Gasnor S.A.

Distribudora de Gas del Centro S.A.

Distribudora de Gas Cuyana S.A.

Camuzzi Gas del Sur S.A.

Gasnea S.A.

Litoral Gas S.A.

Metrogas S.A.

Gas Natural Ban S.A.

Camuzzi Gas Pampeana S.A.

BOLIVIA

BRAZIL

PARAGUAY

Jujuy

Salta

Formosa

Asunción

Chaco

Tucumán

Catamarca

Santiago del Estero

Misiones

La Rioja

Corrientes

San Juan

Santa Fe

Córdoba

Entre Ríos

CHILE

Santiago

URUGUAY

San Luis

Buenos Aires

Mendoza

Montevideo

La Pampa

**ARGENTINA**

Buenos Aires

Neuquén

Río Negro

Chubut

South Pacific Ocean

Santa Cruz

South Atlantic Ocean

N

Km

0    250    500

——— Tranportadora de Gas del Norte S.A.

——— Tranportadora de Gas del Sur S.A.

*Source: Enargas.*

**Map 8** Argentina's cross-border gas links with neighbouring countries, 2002

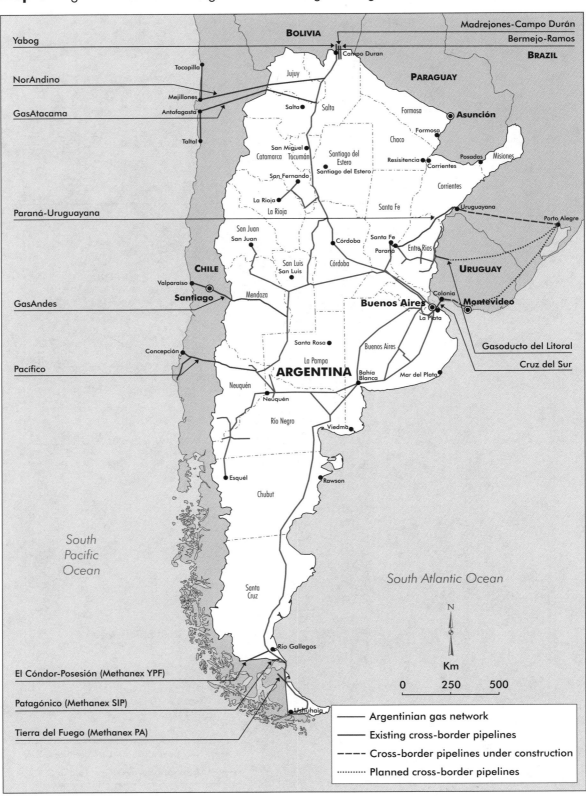

Sources: Enargas, IEA.

**Map 9** Bolivia's natural gas basins and transmission infrastructure, 2002

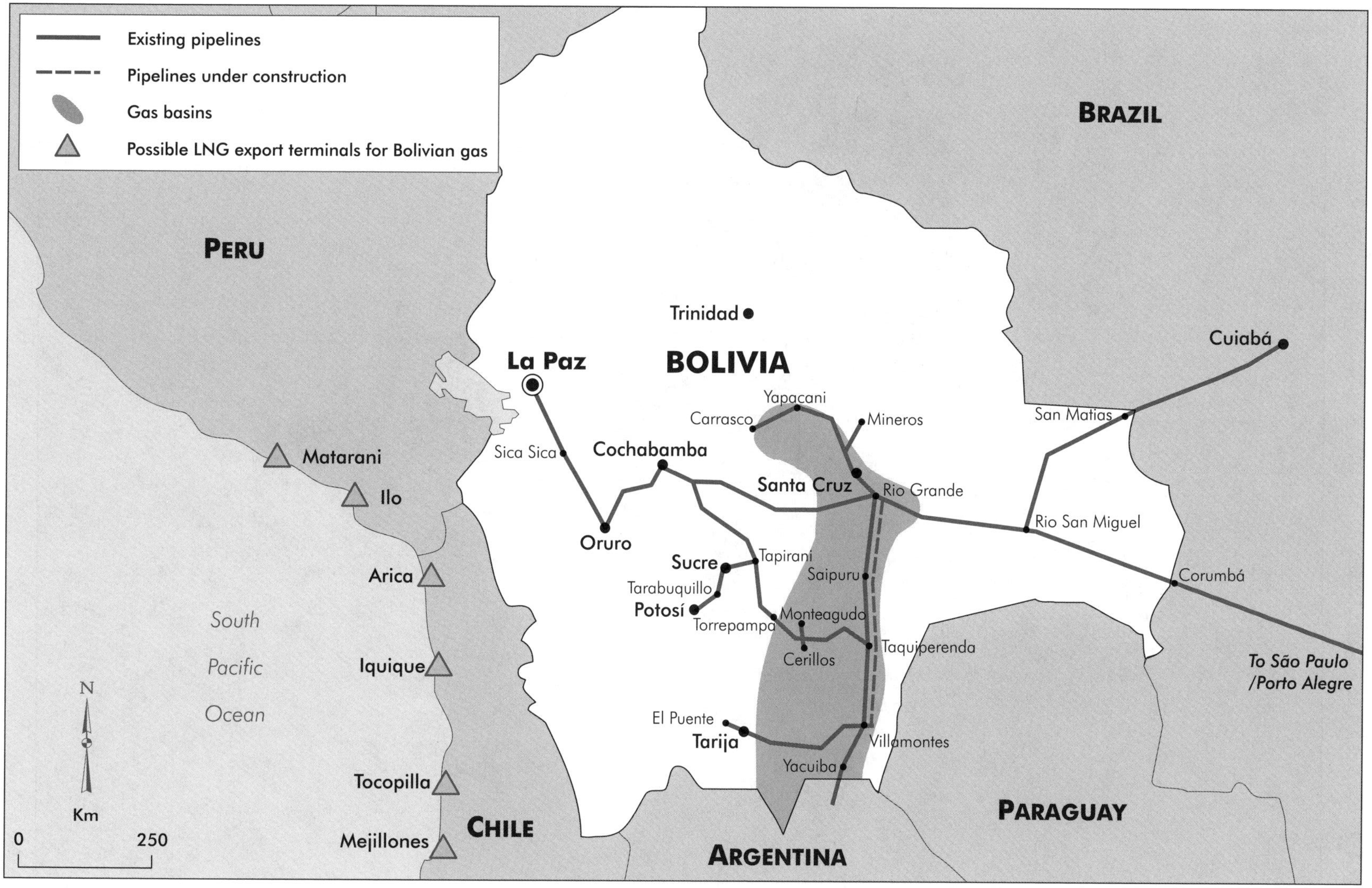

Legend:
- Existing pipelines
- Pipelines under construction
- Gas basins
- Possible LNG export terminals for Bolivian gas

BRAZIL

Cuiabá

To São Paulo /Porto Alegre

Corumbá

San Matías

Rio San Miguel

PARAGUAY

Mineros

Rio Grande

Taquiperenda

Yapacani

Santa Cruz

Saipuru

Villamontes

Trinidad

BOLIVIA

Carrasco

Tapirani

Monteagudo

Cerillos

Yacuiba

Cochabamba

Tarabuquillo

Torrepampa

El Puente

Tarija

Sucre

Potosí

ARGENTINA

Oruro

La Paz

Sica Sica

PERU

Matarani

Ilo

Arica

Iquique

Tocopilla

Mejillones

CHILE

South Pacific Ocean

N

0    250    Km

Source: IEA.

## Map 10 Brazil's regions and states

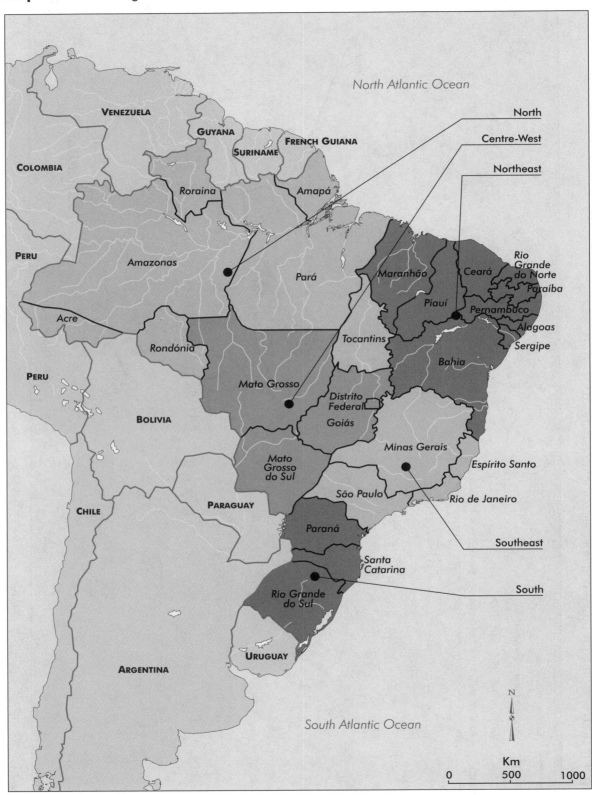

**Map 11** Brazil's natural gas reserves and transmission network, 2002

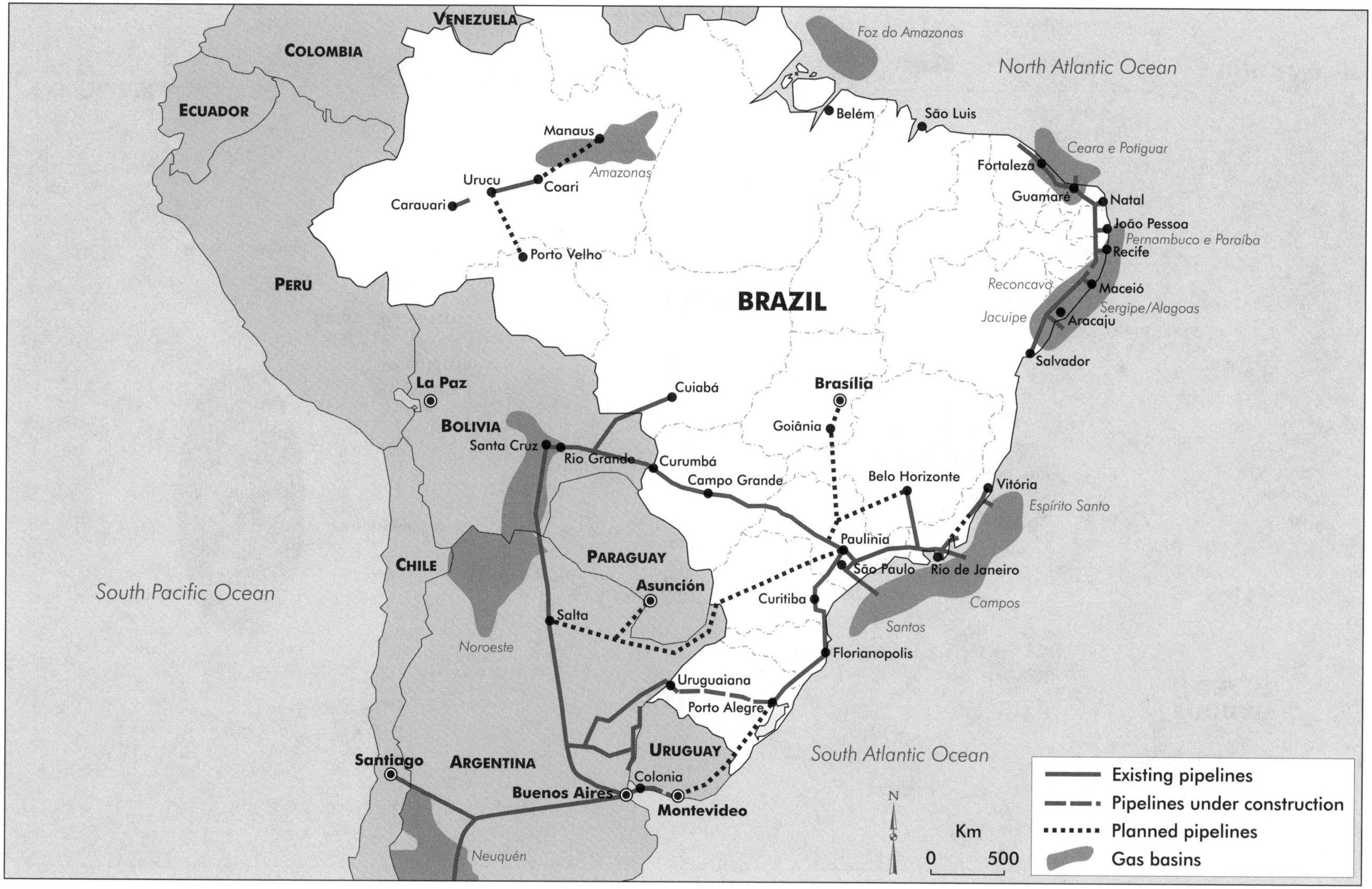

Legend:
— Existing pipelines
—·— Pipelines under construction
······ Planned pipelines
Gas basins

Km
0   500

Source: IEA.

**Map 12** Venezuela's natural gas basins

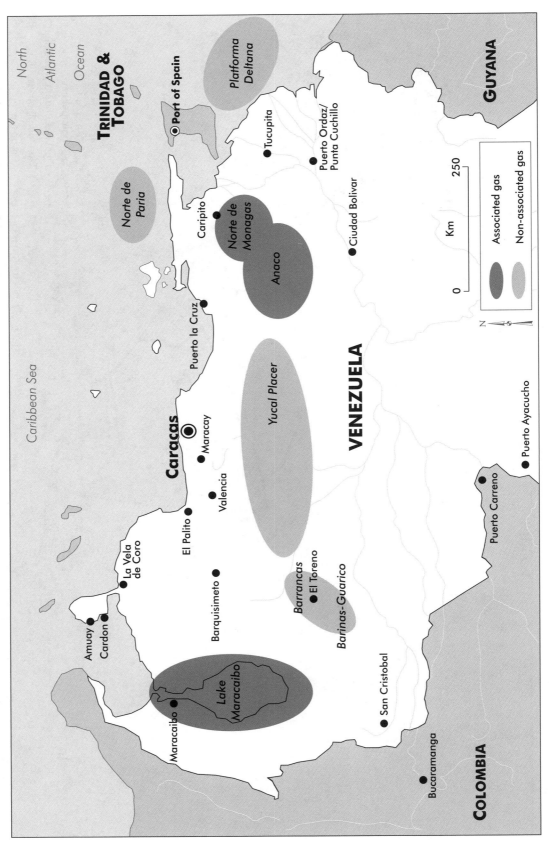

*Source: PDVSA.*

**Map 13** Venezuela's natural gas transmission infrastructure, 2001

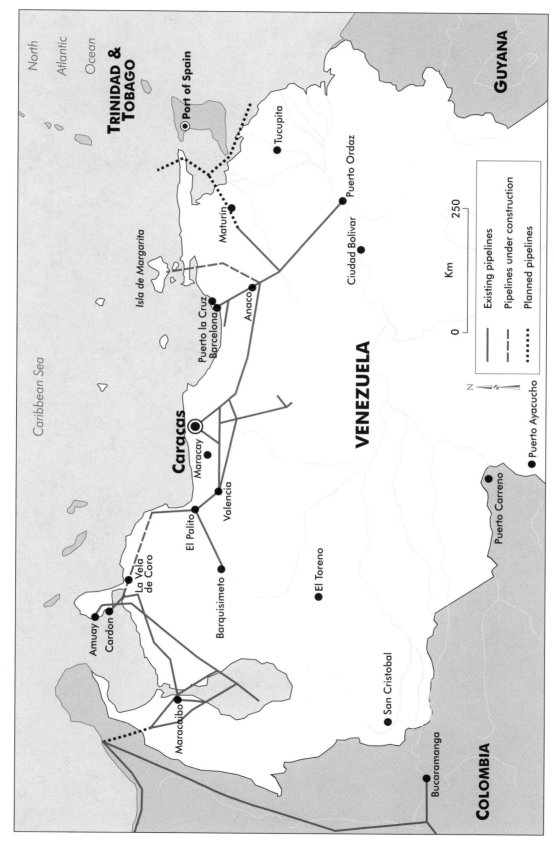

*Source: IEA.*

# ORDER FORM

## IEA BOOKS

**Fax: +33 (0)1 40 57 65 59**
**E-mail: books@iea.org**
**www.iea.org/books**

**INTERNATIONAL ENERGY AGENCY**

9, rue de la Fédération
F-75739 Paris Cedex 15

## *I would like to order the following publications*

| PUBLICATIONS | ISBN | QTY | PRICE | TOTAL |
|---|---|---|---|---|
| ☐ **South American Gas - *Daring to Tap the Bounty*** | **92-64-19663-3** | | **€100** | |
| ☐ World Energy Outlook 2002 | 92-64-19835-0 | | €150 | |
| ☐ Energy Policies of IEA Countries – 2002 Review (Compendium) | 92-64-19773-7 | | €120 | |
| ☐ Flexibility in Natural Gas Supply and Demand | 92-64-19938-1 | | €75 | |
| ☐ Natural Gas Information 2002 | 92-64-19791-5 | | €150 | |
| ☐ Developing China's Natural Gas Market - *The Energy Policy Challenges* | 92-64-19837-7 | | €100 | |
| ☐ Security of Supply in Electricity Markets - *Evidence and Policy Issues* | 92-64-19805-9 | | €100 | |
| ☐ Distributed Generation in Liberalised Electricity Markets | 92-64-19802-4 | | €75 | |
| | | **TOTAL** | | |

## DELIVERY DETAILS

Name _____ Organisation _____

Address _____

_____

Country _____ Postcode _____

Telephone _____ E-mail _____

## PAYMENT DETAILS

☐ I enclose a cheque payable to IEA Publications for the sum of $_____ or €_____

☐ Please debit my credit card (tick choice). ☐ Access/Mastercard ☐ VISA ☐ American Express

Card no: ⌷⌷⌷⌷⌷⌷⌷⌷⌷⌷⌷⌷⌷⌷⌷⌷⌷⌷⌷

Expiry date: ⌷⌷⌷⌷⌷⌷ Signature:

**OECD PARIS CENTRE**
Tel: +33 (0)1 45 24 81 67
Fax: +33 (0)1 49 10 42 76
E-mail: distribution@oecd.org

**OECD BONN CENTRE**
Tel: +49 (228) 959 12 15
Fax: +49 (228) 959 12 18
E-mail: bonn.contact@oecd.org

**OECD MEXICO CENTRE**
Tel: +52 (5) 280 12 09
Fax: +52 (5) 280 04 80
E-mail: mexico.contact@oecd.org

*You can also send your order to your nearest OECD sales point or through the OECD online services:*
*www.oecd.org/ bookshop*

**OECD TOKYO CENTRE**
Tel: +81 (3) 3586 2016
Fax: +81 (3) 3584 7929
E-mail: center@oecdtokyo.org

**OECD WASHINGTON CENTER**
Tel: +1 (202) 785-6323
Toll-free number for orders:
+1 (800) 456-6323
Fax: +1 (202) 785-0350
E-mail: washington.contact@oecd.org

International Energy Agency, 9 rue de la Fédération, 75739 Paris CEDEX 15
PRINTED IN FRANCE BY CHIRAT
(61 03 01 1 P1) ISBN 92-64-19663-3 - 2003